Progress in Inflammation Research

Series Editor

Prof. Michael J. Parnham, PhD
Senior Scientific Advisor
PLIVA Research Institute Ltd.
Prilaz baruna Filipovića 29
HR-10000 Zagreb
Croatia

Advisory Board

Forthcoming titles:

Toll-like Receptors in Inflammation, L.A.J. O'Neill, E. Brint (Editors), 2006

Chemokine Biology: Basic Research and Clinical Application, Volume I: Immunobiology of Chemokines, K. Neote, L. G. Letts, B. Moser (Editors), 2005

Chemokine Biology: Basic Research and Clinical Application, Volume II: Pathophysiology of Chemokines, K. Neote, L. G. Letts, B. Moser (Editors), 2005

The Hereditary Basis of Rheumatic Diseases, R. Holmdahl (Editor), 2006

(Already published titles see last page.)

Complement and Kidney Disease

Peter F. Zipfel

Editor

Birkhäuser Verlag
Basel · Boston · Berlin

Editor

Peter F. Zipfel
Department of Infection Biology
Leibniz Institute for Natural Product Research and Infection Biology
Hans Knoell Institute
Beutenbergstr. 11a
D-07745 Jena
Germany

Library of Congress Cataloging-in-Publication Data

Complement and kidney disease / Peter F. Zipfel, editor.
p. ; cm. -- (Progress in inflammation research)
Includes bibliographical references and index.
ISBN-13: 978-3-7643-7166-1 (alk. paper)
ISBN-10: 3-7643-7166-8 (alk. paper)
1. Kidneys--Diseases--Etiology. 2. Complement (Immunology). I. Zipfel, Peter F. II. PIR (Series)
[DNLM: 1. Kidney Diseases--etiology. 2. Complement--physiology.
3. Hemolytic-Uremic Syndrome. WJ 300 C7373 2005]
RC903.C63 2005
61636'1--dc22

2005053631

Bibliographic information published by Die Deutsche Bibliothek
Die Deutsche Bibliothek lists this publication in the Deutsche Nationalbibliografie;
detailed bibliographic data is available in the internet at http://dnb.ddb.de

ISBN-10: 3-7643-7166-8 Birkhäuser Verlag, Basel – Boston – Berlin
ISBN-13: 978-3-7643-7166-1 Birkhäuser Verlag, Basel – Boston – Berlin

Part of Springer Science+Business Media
Printed on acid-free paper produced from chlorine-free pulp. TCF ∞
Cover design: Markus Etterich, Basel
Cover illustration: HUS – Renal biopsy with typical glomerular changes of HUS including segmental thickening of glomerular basement membranes with double contours, congestion of capillaries and increase in extracellular matrix (Masson trichrome stain) (see p. 173)
Printed in Germany
ISBN-10: 3-7643-7166-8 e-ISBN: 3-7643-7428-4
ISBN-13: 978-3-7643-7166-1

9 8 7 6 5 4 3 2 1 www.birkhauser.ch

Contents

Abbreviations

ADAMTS13 an integrin-like and metalloproteinase with thrombospondin motif
APC antigen presenting cells
BSH British Society for Haematology
CCP complement control protein module also termed short consensus repeat (SCR)
CCE cholesterol crystal emboli
CR1 complement receptor 1, also termed CD 36
DAF decay accelerating Factor, also termed CD 59
DDD dense deposit diseases
DHPLC denaturing high performance liquid chromatography
ESRD end stage renal disease
I/R ischemic reperfusion injury
GBM glomerular basement membrane
HUS hemolytic uremic syndrome
mAB monoclonal antibody
MAC membrane attack complex, the terminal effector system of complement which generates holes in the membrane of the target cell
MCP membrane cofactor protein also termed CD 46
MN membranous nephropathy
MPGN membranoproliferative glomerulonephritis
MLPA multiplex ligation-dependent probe amplification
RCA regulators of complement gene cluster on human chromosome 1, includes the genes coding for Factor H, MCP, CR1 and DAF
PE plasma exchange
PBMC peripheral blood mononuclear cells
SLE systemic lupus erythematosus
Stx Shiga like toxin
SSCP single strand conformation polymorphism
TTP thrombotic thrombocytic purpura
TMA thrombotic microangiopathies
vWF von Willebrand Factor

List of contributors

John P. Atkinson, Division of Rheumatology, Washington University School of Medicine, St. Louis, MO 63110, USA; e-mail: jatkinso@im.wustl.edu

Stefan P. Berger, Department of Nephrology, C3-P25, Leiden University Medical Center, P.O. Box 9600, 2300 RC Leiden, The Netherlands; e-mail: s.p.berger@lumc.nl

Markus Cybulla, Department of Nephrology, Albert-Ludwigs-University, Hugstetter Straße 55, 79106 Freiburg, Germany;
e-mail: markus.cybulla@uniklinik-freiburg.de

Mohamed R. Daha, Department of Nephrology, C3-P25, Leiden University Medical Center, P.O. Box 9600, 2300 RC Leiden, The Netherlands;
e-mail: m.r.daha@lumc.nl

Véronique Frémeaux-Bacchi, Hôpital Européen Georges Pompidou, 20 rue Leblanc, 75908 Paris Cedex 15, France;
e-mail: veronique.fremeaux-bacchi@hop.egp.ap-hop-paris.fr

Tim Goodship, Institute of Human Genetics, University of Newcastle upon Tyne, Newcastle upon Tyne, NE1 3BZ, United Kingdom; e-mail: t.h.j.goodship@ncl.ac.uk

T.W.L. Groeneveld, Department of Nephrology, C3-P25, Leiden University Medical Center, P.O. Box 9600, 2300 RC Leiden, The Netherlands;
e-mail: t.w.l.groeneveld@lumc.nl

Stefan Heinen, Department of Infection Biology, Leibniz Institute for Natural Product Research and Infection Biology, Hans Knoell Institute, Beutenbergstr. 11a, 07745 Jena, Germany; e-mail: StHeinen@hki-jena.de

V. Michael Holers, Division of Rheumatology, B-115, University of Colorado, Health Sciences Center, 4200 East 9th Avenue, Denver Co 80262, USA;
e-mail: Michael.Holers@UCHSC.edu

Michael Hoffmann, Department of Laboratory Medicine, Albert-Ludwigs-University, Hugstetter Str. 55, 79106 Freiburg, Germany;
e-mail: michael.markus.hoffmann@uniklinik-freiburg.de

Bernd Hoppe, Children's Hospital of the University of Cologne, Pediatric Nephrology, Kerpener Str. 62, 50937 Cologne, Germany;
e-mail: bernd.hoppe@uk-koeln.de

Mihály Józsi, Department of Infection Biology, Leibniz Institute for Natural Product Research and Infection Biology, Hans Knoell Institute, Beutenbergstr. 11a, 07745 Jena, Germany; e-mail: Mihaly.Jozsi@hki-jena.de

Pearl L. Lewis, Maryland Patient Advocacy Group, Foundation for Children with Atypical HUS, Maryland Renal Advocate 9131, 2530 Kensington Gardens #104, Ellicott City, MD 21043, USA; e-mail: yettapearl@comcast.net

Peter Lachmann, Centre for Veterinary Science, Madingley Road, Cambridge, CB3 0ES, UK; e-mail: pjl1000@cam.ac.uk

Christoph Licht, Children's Hospital of the University of Cologne, Pediatric Nephrology, Kerpener Str. 62, 50937 Cologne, Germany;
e-mail: christoph.licht@uni-koeln.de

Momir Macanovic, Centre for Veterinary Science, Madingley Road, Cambridge, CB3 0ES, UK; e-mail: momirmacanovic@hotmail.com

Hartmut P.H. Neumann, Department of Nephrology, Albert-Ludwigs-University, Hugstetter Str. 55, 79106 Freiburg, Germany;
e-mail: neumann@med1.ukl.uni-freiburg.de

Marina Noris, Chem Pharm D, Transplant Research Center "Chiara Cucchi de Alessandri e Gilberto Crespi", Mario Negri Institute for Pharmacological Research, Villa Camozzi, Via Camozzi 3, 24020, Ranica (BG), Italy;
e-mail: noris@marionegri.it

Giuseppe Remuzzi, Chem Pharm D, Transplant Research Center "Chiara Cucchi de Alessandri e Gilberto Crespi", Mario Negri Institute for Pharmacological Research, Villa Camozzi, Via Camozzi 3, 24020, Ranica (BG), Italy; and Department of Medicine and Transplantation, Ospedali Riuniti di Bergamo, Largo Barozzi 1, 24128, Bergamo, Italy; e-mail: gremuzzi@marionegri.it

Peter Riegler, Department of Nephrology, General Hospital, Lorenz-Böhler-Str. 5, 39100 Bolzano, Italy; e-mail: Peter.Riegler@asbz.it

Anja Roos, of Nephrology, C3-P25, Leiden University Medical Center, P.O. Box 9600, 2300 RC Leiden, The Netherlands; e-mail: a.roos@lumc.nl

Steven H. Sacks, Department of Nephrology and Transplantation, Guy's, King's and St. Thomas' School of Medicine, King's College of London, London SE1 9RT, UK; e-mail: steven.sacks@kcl.ac.uk

Maren Salzmann, Department of Nephrology, Albert-Ludwigs-University, Hugstetter Straße 55, 79106 Freiburg, Germany; e-mail: msalzmann@web.de

Gisa Schluh, Department of Nephrology, Albert-Ludwigs-University, Hugstetter Straße 55, 79106 Freiburg, Germany; e-mail: schluh@medizin.ukl.uni-freiburg.de

Christine Skerka, Department of Infection Biology, Leibniz Institute for Natural Product Research and Infection Biology, Hans Knoell Institute, Beutenbergstr. 11a, 07745 Jena, Germany; e-mail: christine.skerka@hki-jena.de

Richard J.H. Smith, Department of Otolaryngology, 200 Hawkins Drive, 21151 PFP, The University of Iowa, Iowa City, IA 52242, USA; e-mail: richard-smith@uiowa.edu

Joshua M. Thurman, Division of Nephrology and Hypertension, B-115, University of Colorado Health Sciences Center, 4200 East 9th Avenue, Denver, CO 80262, USA; e-mail: joshua.thurman@uchsc.edu

Reinhard Würzner, Department of Paediatrics, Innsbruck Medical University, Anichstr. 35, 6020 Innsbruck, Austria

Wuding Zhou, Department of Nephrology and Transplantation, Guy's, King's and St. Thomas' School of Medicine, King's College of London, London SE1 9RT, UK; e-mail: wuding.zhou@kcl.ac.uk

Lothar Bernd Zimmerhackl, Department of Paediatrics, Innsbruck Medical University, Anichstr. 35, 6020 Innsbruck, Austria; e-mail: lothar-bernd.zimmerhackl@uklibk.ac.at

Peter F. Zipfel, Department of Infection Biology, Leibniz Institute for Natural Product Research and Infection Biology, Hans Knoell Institute, Beutenbergstr. 11a, 07745 Jena, Germany, and Friedrich-Schiller-University Jena, Professor for Infection Biology; e-mail: peter.zipfel@hki-jena.de

In Dämmerschein liegt schon die Welt erschlossen,
Der Wald ertönt von tausendstimmigem Leben,
Tal aus, Tal ein ist Nebelstreif ergossen,
Doch senkt sich Himmelsklarheit in die Tiefen,...

Faust II, Johann Wolfgang v. Goethe

Preface

Complement is a central and important immune defense system that acts at the interface between innate and adaptive immunity. Aimed at maintaining the integrity of the human organism, this system is – when activated – highly toxic, destructive and inflammatory. Actually, the action of the highly efficient and toxic activation components is desired and favorable on foreign surfaces such as microbes, or surfaces of modified self cells, but at the same time absolutely unfavorable on the surface of host cells. Consequently, activation of this defense system must be restricted in terms of time and space. Complement activation is initiated either spontaneously by the alternative pathway, or is induced by antibodies via the classical pathway. Regardless of the pathway of initiation, in order to prevent attack and damage host cells and tissues must continuously restrict and block amplification of the complement cascade. Uncontrolled activation and defective control results in autoimmune diseases.

It has been known for a long time that there is an association between the complement system and renal disease. Complement components are involved in the initiation, pathophysiology and progression of glomerular diseases in many ways. The recent years have witnessed an impressive development in defining the role of complement and individual complement components in glomerular disease. In several cases the defects have been identified on a molecular level by describing novel mutations in individual complement components, and by understanding the pathophysiology of the diseases, e.g., by defining how mutations of single components relate to glomerular damage. This information reveals a pivotal role of complement in renal disease and – in several cases – allowed a novel understanding of the disease mechanisms or a description of the pathophysiology in molecular terms. In addition to reports on the general role of complement in glomerular disease, this volume highlights the role of individual complement components in kidney diseases. In particular, several kidney diseases such as the atypical form of hemolytic uremic syndrome, membranoproliferative glomerulonephritis and systemic lupus erythematosus are covered in detail from the clinical side and from the side of basic research.

This book is aimed at bridging the broad scope between basic research, disease mechanisms and clinical aspects, and also include a report from the patients' side. Several distinguished researchers, clinicians and a patient advocate, who have made significant contributions to the field of complement and glomerular disease during the last years review the latest findings in this field and give an in depth overview on the current situation. The understanding of the relation of complement and glomerular diseases has substantially improved in recent years, due to the reports of gene mutations in single effector proteins or regulators, detailed functional characterization of single mutated complement components and the description how defective function relates to disease. Through these various concepts, novel diagnostic regimens have been established and new ways paved for therapeutic intervention and treatment.

Based on the broad range of topics reviewed and covered, this book is aimed for anyone with an interest in glomerular diseases and in the role of complement in autoimmunity, particularly clinicians, researchers and students who wish to understand the role of complement in the pathogeneses of renal and autoimmune diseases.

Jena, September 2005 Peter F. Zipfel

The complement system in renal diseases

Momir Macanovic and Peter Lachmann

Centre for Veterinary Science, Madingley Road, Cambridge, CB3 0ES, UK

Introduction

The association between the complement system and renal disease has been appreciated for a long time. The complement system may be involved in the initiation, pathogenesis and resolution of glomerulonephritis (GN) in many ways. The activation of complement and its deleterious consequences have been observed in many renal diseases: some primary forms of GN, GN associated with systemic diseases, acute and chronic humoral rejection of renal transplant, cholesterol renal emboli, hemodialysis and others.

Understanding of relations of complement and GN evolved from the initial observation of depletion of serum complement in some forms of GN and identification of complement deposits in kidney biopsy specimens. Varying glomerulonephritides are associated with alterations in the serum concentration of specific complement components, the presence of complement breakdown products in the circulation, glomerular complement deposits and circulating complement-activating factors.

The evidence for the role of complement in enhancing renal injury comes also from experimental data obtained by depletion of complement with cobra venom. Studies utilizing cobra venom factor to produce generalized complement depletion in animals have shown that most of the morphological and functional changes which occur in some forms of GN are complement dependent [1]. More recently, molecular biology techniques of genetically manipulated mice, administration of recombinant complement regulatory proteins [2, 3], monoclonal anti-C5 antibodies that block or reduce kidney damage [4–6], have provided additional evidence for multiple and important roles of complement in glomerular diseases and implications for complement-targeted therapy to control ongoing inflammation in the kidney.

Hypocomplementemia in GN

Evidence of complement activation in GN comes from characteristic patterns of a decrease in the serum concentrations of specific components, some of which are vir-

Complement and Kidney Disease, edited by Peter F. Zipfel

tually diagnostic of certain nephritides. Hypocomplemententemia is used as a marker for diagnosis and monitoring efficacy of treatment. The components most frequently measured in clinical practice are C3, C4, C2 and total hemolytic complement (CH50). Several commercial antisera are available for each, and rapid nephelometric assays are in widespread use. The CH50 is a functional measurement, based on the ability of serum to support complement-mediated lysis of sensitized erythrocytes.

In chronic bacteremic states or in lupus nephritis, circulating or deposited immune complexes may act as classical pathway activators [7]. These disorders are frequently accompanied by reduction of the classical pathway activation proteins C1q, C2, C4 along with C3 [8, 9]. Typical patients with acute post-streptococcal GN have reductions of C3, C5 and properdin. Normalization of C3 usually occurs within 6–8 weeks [10].

The other major group of hypocomplementic GN, membranoproliferative GN (MPGN), is somewhat more variable. Type II MPGN usually is associated with a reduction in C3 only. Evidence of either classical pathway activation or terminal component depression is seen in type I MPGN.

In all these situations, depression of serum complement component concentrations is assumed to represent activation, although synthetic defects are occasionally suggested.

Complement activation fragments in circulation

Complement activation, whether in the fluid phase or on surface, often results in the generation of soluble fragments, which can be detected in circulation. The detection of these complement fragments during the course of glomerular injury provides additional evidence of ongoing complement activation. The fragments most often assayed are C3a and C5a. Commercial assays are available for both, and elevated levels appear to be compatible with ongoing complement activation. Neither, however, is currently in wide clinical usage. These tests are frequently used in testing biocompatibility of dialysis membranes.

Complement activation in hemodialysis patients occurs due to the bioincompatibility of some types of dialysis membranes [11]. Complement activation proceeds via the alternate pathway during hemodialysis [12–14]. New cuprophane membranes activate complement to the greatest degree [15]. The hydroxyl group on the surface of the cuprophane membrane is thought to promote the deposition of C3b and the association of C3b with factor B; this is followed by the activation of factor B by factor D, eventually resulting in formation of the C3 convertase C3bBb and the C5 convertase. There are several sequelae of complement activation on hemodialysis membranes: release of anaphylatoxins (C3a and C5a), formation of the membrane attack complex (C5b-9) and activation of neutrophils and monocytes. C3a and C5a are potent, biologically active agents capable of producing intense vascular smooth muscle contraction, increased vascular permeability, and the release of histamines from mast cells [16].

Several types of membranes result in specific patterns of complement activation. Complement activation is greatest with cuprophane, intermediate with cellulose acetate and minimal with polyacrylonitrile [17].

Complement deposits in glomeruli

Deposition of complement, visualized by immunofluorescence, immunochemistry or electron microscopy is a frequent feature of GN, usually in parallel with antibody deposition [2, 8, 18]. Complement deposits may be dominant, for example in MPGN, or characteristic such as in post-streptococcal GN, IgA nephropathy, and membranous nephropathy (MN). In class IV lupus nephritis and some forms of rapidly progressive GN, immune deposits including complement are generally found in mesangial and subendothelial distribution. Complement deposits generally contain either both C4 and C3, corresponding to the plasma patterns of classical complement activation, or C3 without the early components, corresponding to alternative pathway of complement activation [19].

Immune complex diseases and complement

Multiple relationships exist between complement system and immune-mediated nephropathies. Most forms of glomerulonephritides are associated with immune complex deposition. Under physiological conditions, complement promotes the clearance of immune complexes, both modifying the immune complex size and favoring the physiological clearance by the erythrocyte transport system [20–22]. Deposition of complement proteins (C3bi) on the surface of immune complexes facilitates clearance through interaction with erythrocyte-bound CR1 and the reticuloendothelial system.

However, depending on circumstances, complement may be not a friend, it can be a foe. If immune complexes cannot be eliminated, then complement becomes chronically activated and can incite inflammation. Chronic infections can perpetuate the formation of immune complexes, which in hepatitis C infection and bacterial endocarditis cause relentless activation and consumption of complement.

Immune complexes in glomeruli, either deposited from the circulation or generated *in situ*, can activate the complement system. Active products of this system include the anaphylatoxins C3a and C5a, C3b, and the C5b-9 membrane attack complex. The outcomes of this are glomerular changes: either increased permeability of glomerular basement membrane (GBM) or inflammation. The pattern of glomerular injury seen in immune complex-mediated glomerular diseases is related to the site of formation of immune deposits. If complement is activated at a site accessible to blood constituents, such as subendothelial and mesangial regions, the generated C3a, C5a and C3b can interact with circulating or intrinsic glomerular cells bearing relevant receptors, and striking changes seen by light microscopy occur, particularly hypercellularity, which may reflect infiltration by circulating inflammatory cells such as neutrophils and monocytes, or proliferation of glomerular cells, particularly mesangial cells. The histological result is proliferative GN (focal proliferative, diffuse proliferative, proliferative and exudative, membranoproliferative, rapidly progressive) and clinically active urine sediment (red cells, white cells, and cellular and granular casts), proteinuria, and often an acute decline in renal functions. The examples are post-infectious GN, IgA nephropathy, rapidly progressive GN, lupus nephritis, and MPGN.

In a privileged site such as subepithelial space, immune deposits can also activate complement, but there is no influx of inflammatory cells, since the chemoattractants are separated from the circulation by the GBM. Thus, injury is limited to the GBM and glomerular epithelial cells and the primary clinical manifestation is proteinuria, which is often in the nephrotic range. Histologically, these patients most commonly have MN [3, 21, 22].

A plethora of recent studies establishes that both circulating inflammatory cells and resident glomerular cells can mediate glomerular injury acutely by release of oxidants, proteases and probably other chemoattractants and GBM-degrading molecules. Chronic injury is also augmented by release of various cytokines and growth factors, which results in increased deposition of extracellular matrix leading to scarring and sclerosis.

Deficiencies of complement components and GN

Deficiencies or polymorphism of certain complement components are also associated with disease, especially with infections but also with autoimmune and renal diseases. The detailed description of this topic is presented in separate chapters of this book (see chapters by Goodship et al., Skerka and Józsi, and Zipfel et al.).

Our understanding of some of the roles of complement derived from the studies of individuals or experimental animals deficient in some of complement components. Several of these deficiencies are associated with renal disease, either in primary renal diseases or in systemic disease with renal involvement, such as systemic lupus erythematosus (SLE). Complement deficiencies can cause renal disease by uncontrolled complement activation, or aberrant handling of immune complexes or secondary to the development of SLE.

Primary C3 deficiencies result in increased susceptibility to infections and in MPGN [21–24]. The renal disease in these patients may be explained by an aberrant handling of immune complexes in the absence of C3. A rare relation between C9 deficiency and renal disease has been described [3]. One report relates C9 deficiency to immune complex GN [25], and another report to IgA nephropathy [26]. How C9 deficiency contributes to renal disease is unknown. Mice deficient in clusterin, a fluid-phase complement regulator, which inhibits the incorporation of C5b-9, also develop renal disease [27].

A direct interaction between complement and renal injury was concluded from families with a Factor H deficiency [3, 28, 29]. Factor H deficiency results in low plasma levels of Factor B, unhindered activation of fluid-phase C3, severe depletion of plasma C3 and in consumption of the terminal complement components C5–C9 (see chapter by Zipfel et al.). Continuous activation and turnover of C3 in the vicinity of the GBM may cause C3b to bind to glomeruli and incite inflammation.

Both MPGN and idiopathic hemolytic uremic syndrome are associated with Factor H deficiency (see chapters by Skerka and Józsi, and Zipfel et al.). A high index of suspicion for Factor H deficiency is needed in patients with reduced levels of C3 and recurrent hemolytic–uremic syndrome or MPGN. Factor H-deficient pigs [30, 31] and Factor H knockout mice [32] develop spontaneous renal disease and display MPGN-like symptoms, confirming the importance of Factor H in complement regulation. In pigs, Factor H deficiency leads to the development of MPGN with dense deposits, similar to human type II MPGN observed in patients with nephritic factor (NeF) and in some very rare patients with Factor H deficiency [29]. The administration of purified Factor H to pigs prevents the formation of immune deposits and subsequent glomerular damage, and, even when administered later, allows regression of nephritis. Factor H deficit/mutation carries a great risk (21–65%) of recurrence in the living donor renal transplant (see chapter by Goodship et al.).

Factor I deficiency results in uncontrolled C3 activation and recurrent infections. Some of these patients exhibit focal segmental GN [33], indicating that uncontrolled activation of C3 results in the generation of active C3 fragments and renal disease.

Antibodies against complement components

In autoimmune individuals an antibody response against complement components can occur. Some of these antibodies show such a high degree of correlation with renal disease that the term NeF was introduced to indicate this activity. Anti-complement autoantibodies can contribute to renal disease by deregulating complement activation or influencing immune complex deposition in glomeruli. Several autoantibodies directed against complement components have been identified in patients [3], some of which are related to renal disease directly: C3 nephritic factor (C3NeF), C4NeF, C3NeF:P, anti-Factor H, and anti-C1q.

One of the first demonstrations of complement activation by circulating factor was the recognition that serum of some patients with type II MPGN was capable of activating C3. This activation has been shown to result from an immunoglobulin, C3NeF, which binds to and stabilizes C3bBb, the alternative pathway C3 convertase, and prolongs its half-life, resulting in ongoing complement activation [8, 34–37].

The association of MPGN with disparate disorders such as type II MPGN, partial lipodystrophy, sickle cell disease, complement deficiencies, cryoglobulinemia, and infections with either hepatitis B or C suggests that this disorder is not a single pathogenic entity [18]. The recent recognition of a causal relation between hepatitis C infection and MPGN has led to the suggestion that this virus may be responsible for as many as 60% of cases previously deemed to be idiopathic [38].

C3NeF is an IgG autoantibody that prolongs the enzymatic half-life of the C3 convertase C3bBb, thus producing continuous C3 activation in plasma. Patients with positive C3NeF develop MPGN in 50% of cases. Although clinical manifestations, characterized by heavy proteinuria, progressive loss of renal function, hypertension, are similar, two major types of idiopathic MPGN have been recognized on the basis of difference in ultrastructural morphology: type I, characterized by subendothelial deposits, and type II (dense deposit disease), characterized by the deposition of dense deposits within the GBM.

In type I MPGN immunofluorescence microscopy reveals granular deposits of C3 in the mesangium and in peripheral capillary loops in all patients. Deposits of immunoglobulins and other complement components in capillary loops are present in some patients.

Type II MPGN differs from type I histologically because the dense deposits are localized homogeneously within GBM. C3 and other complement components are detected in the mesangium and capillary loops, but immunoglobulins are generally not seen on immunofluorescence microscopy. Electron microscopy of kidneys affected by type II MPGN reveals electron-dense deposits of unknown composition within the GBM.

Type I MPGN may be mediated by the deposition of immune complexes capable of activating complement both systemically and within the kidney. Serum complement concentrations tend to fluctuate. However, serial determinations reveal at least an intermittent decrease in the concentrations of C3, C1q, and C4 in the vast majority of patients, suggesting activation of complement through both the classic and alternative pathways.

In patients with type II MPGN serum concentrations of C3 tend to be persistently low, while concentrations of early components of the classic pathway are usually normal, suggesting that complement activation occurs primarily through the alternative pathway. Virtually all patients with type II MPGN have high serum concentration of C3NeF [39].

Partial lipodystrophy is a disfiguring condition that affects the body from the waist upward but spares the legs. Mathieson et al. [40] provided an explanation for the loss of fat in this condition. Adipose cells are the main source of Factor D, which completes the formation of the C3 convertase enzyme C3bBb by cleaving Factor B bound to C3b. There is a gradient in the concentration of Factor D in the fat cells of the body; more is present in the upper than the lower half of the body, which could explain the distribution of the fat loss. It is likely that the C3 nephritic antibody in partial lipodystrophy stabilizes the C3bBb C3 convertase that forms in the immediate vicinity of adipocytes. The abnormally stabilized enzyme may then cleave enough C3 to allow assembly of the membrane attack complex, which lyses adipocytes.

C3NeF is not usually connected with other conditions, but it is present in a few patients with SLE [41], post-streptococcal acute GN [42], and even in clinically healthy individual [43]. What triggers the production of C3NeF is unknown. It has been shown that about 50% of the patients positive for NeF do develop MPGN [3, 44, 45].

C3NeF:P has been found in sera from patients with types I and II MPGN and it displays the properties of properdin and IgG. C3NeF:P has been observed in patients with reduced serum concentrations of C3 and terminal complement components [46, 47].

An autoantibody with specificity to classical pathway convertase C4b2a, called C4NeF, stabilizes this convertase and prolongs the half-life [3, 36, 37, 47, 48]. It has been found in the sera of patients with SLE, MPGN and acute post-streptococcal GN. Chronic infections can perpetuate the formation of immune complexes, which in hepatitis C infection and bacterial endocarditis cause relentless activation and consumption of complement.

Defective regulation of C3 typically associated with GN may be due not only to C3NeF, which increases the stability of the C3 convertase enzymes, but also to reduced function of Factor H or Factor I. Other autoantibodies influencing the function of complement regulation have been described in relation to renal disease. Anti-factor H autoantibodies bind to Factor H and inhibit its function of enzymatic inactivation of C3b [49]. The alternative pathway is activated and depleted, leading to secondary C3 deficiency causing MPGN.

Anti-C1q autoantibodies have been described to be related to nephritis in SLE patients [50–52]. About one third of patients with SLE have high titers of autoantibodies against C1q. The presence of these autoantibodies is indicative of severe disease; they are strongly associated with severe consumptive hypocomplementemia and lupus nephritis. A rise in the titer of these anti-C1q autoantibodies has been reported to predict a flare of nephritis [53]. There also seems to be no overt nephritis in anti-C1q-negative SLE patients [54]. Immune deposits eluted from postmortem kidneys of SLE patients reveal the accumulation of these anti-C1q autoantibodies [55]. All these facts point to a pathological role of these autoantibodies. However, not all patients with anti-C1q autoantibodies develop renal disease, and some healthy individuals also have low titer of anti-C1q autoantibodies [56].

Trouw et al. [3] found that anti-C1q antibodies deposit in the glomerulus, together with C1q, after injection of these antibodies in healthy mice, indicating that even in the absence of pre-existing immune complexes, as in SLE, these autoantibodies can target C1q to the glomerulus. The origin of these autoantibodies is unknown, but if C1q forms a molecular association with tissue debris, then it may itself become part of an autoantigenic complex [57].

Complement in SLE and lupus nephritis

There are many other observations that suggest the important role of complement in SLE. In human lupus nephritis, there is a large amount and diversity of immune reactants in glomerular deposits including IgG, IgA, C1q, C4, C3, C5b-9 ("full house" pattern). In addition, there is systemic consumption of complement with reduced concentrations of some components and the appearance of complement activation products in sera. Low serum concentrations of C3 and C4, low total hemolytic complement activity and elevated levels of antibodies to DNA or antinuclear antibody have been reported to correlate with the presence of active GN; serological evidence of increasing disease activity may precede the development of serious renal inflammation by months.

In addition, a number of mouse strains spontaneously develop lupus nephritis with immune complex and complement deposition in glomeruli, similar to the human disease, such as the New Zealand Black/White (NZB/W F1) and MRL/lpr strains.

Deficiencies of complement components are associated with renal disease, secondary to the development of SLE [58]. Deficiency of C1q, C4 and C2 predisposes strongly to the development of SLE via mechanisms relating to defective clearance of apoptotic material. In many of these SLE patients lupus nephritis occurs, characterized by the deposition of immune complexes. These immune complexes are primarily composed of anti-DNA antibodies and nucleosomes as antigen [59].

It has been widely accepted that the activation of complement by immune complexes is an important contributor to tissue injury in patients with SLE. The strength of the association of complement deficiency with SLE itself and with the severity of the disease is inversely correlated with the position of the deficient protein in the activation sequence of the classical pathway. Thus, hereditary homozygous deficiencies of C1q, C1r and C1s, and C4 are each strongly associated with susceptibility to SLE, with respective prevalence of 93%, 57% (since deficiencies of C1r and C1s are usually inherited together), and 75%. By contrast, the prevalence of SLE among persons with C2 deficiency is about 10%.

There is also an association between SLE, hereditary angioedema and GN [60, 61]. In patients with hereditary angioedema, excessive cleavage of C4 and C2 by C1s, caused by a heterozygous deficiency of C1 inhibitor, leads to an acquired deficiency of C4 and C2 that is sufficient to increase susceptibility to SLE and lupus nephritis.

IgA nephropathy and anti-mesangial cell proliferation

In IgA nephopathy serum complement levels are normal. On immunofluorescence of kidney biopsies, in addition to typical mesangial IgA deposits, C3 and C5b-9 in predominantly mesangial distribution are present, whereas C1 and C4 are uncommon, suggesting that IgA-mediated activation of the alternative complement pathway may be involved in the pathogenesis of this form of mesangioproliferative GN [62].

Much attention in recent years has focused on the role of the mesangial cell mediation of immune types of glomerular injury. Mesangial cell proliferation is a prominent feature of glomerular disease, including IgA nephropathy, lupus nephritis, some types of steroid-resistant nephritic syndrome and other lesions. *In vitro* studies demonstrate activation of mesangial cells by C5b-C9 [20].

Complement fixing IgG or IgA antibodies to mesangial cells have been reported in IgA nephropathy [63]. IgA can activate complement by the alternative pathway [64], and C5b-9 deposits are prominent in idiopathic IgA GN [65].

To test the role of C5-9 in immune diseases of the mesangium, anti-thymocyte serum model of mesangioproliferative GN was induced in C-6-deficient rats. There was a marked reduction in glomerular mesangiolysis, platelet infiltration, mesangial cell proliferation, macrophage infiltration and matrix expansion in C-6-deficient rats compared to control animals [66].

Nangaku et al. [67] compared a model of antibody to glomerular endothelial cell-induced thrombotic microangiopathy and found marked reduction in intracapillary thrombi and fibrin deposits, glomerular endothelial cell proliferation and macrophage infiltration in C6-deficient animals.

GN associated with infection

The prototype of this form of GN is the nephritis that follows infection with nephritogenic strains of group-A hemolytic streptococci by 14–21 days. Complement abnormalities include a large reduction in CH50 and C3 concentrations in many cases with normal C4, suggesting complement activation primarily via the alternative pathway [68]. Coarse granular pattern of deposits of C3 are seen by immunofluorescence in the mesangium and along capillary walls accompanied by lesser amounts of IgG, and suggest that immune complex formation (either circulating or formed *in situ*) is involved [38].

Subendothelial immune deposits are probably responsible for local influx of inflammatory cells, but they are rapidly cleared and may not be seen on renal biopsy specimens obtained relatively late in the course of disease. Large subepithelial immune deposits referred as "humps" are best seen on electron microscopy during the first 2 weeks of the disease and tend to diminish by weeks 4–8.

Serial measurements of complement components can be helpful in the diagnosis of this disorder. Total hemolytic complement activity and C3 concentration are depressed early in the course of the disease and, in most cases, return to normal by 6–8 weeks [69]. The finding of persistently low concentrations of C3 more than 8 weeks after presentation should alert clinician to the possibility of lupus nephritis or MPGN.

Membranous nephropathy

MN disease is mediated primarily by the humoral immune response, which leads to deposition of IgG and complement on the outer, subepithelial surface of the GBM [70]. Although small complexes can cross the GBM and deposit at this site, experimental studies suggest that passive glomerular trapping of preformed immune complexes directly from the circulation is unlikely. Deposits are formed *in situ* by accretion of an antibody to intrinsic or planted antigens on the epithelial side of GBM.

Based on studies in animal models, the mechanism by which damage to the glomerular filtration barrier occurs that is sufficient to cause proteinuria appears to involve sublytic effects of complement C5b-9 on the glomerular epithelial cell. Complement activation and cleavage of C5 generates chemotactic factor C5a, which presumably is flushed by filtration forces into the urinary space and does not move backwards across the GBM to attract circulating inflammatory cells. The other product of C5 cleavage C5b combines with C6 to form a lipophilic complex that inserts into the lipid bilayer of the glomerular epithelial cells where C7, C8 and multiple C9 molecules are added to create a pore-forming complex C5b-9 [20].

Because the deposits form only on the outer, or subepithelial surface of the glomerular capillary wall, complement and immunoglobulin-derived chemotactic and immune adherence proteins are nor interactive with circulating cells, probably accounting for the non-inflammatory nature of the lesion. However, proteinuria is complement dependent and appears to be mediated primarily by the C5b-9 membrane attack complex of complement. In addition to other proteins, urinary excretion of C5-9 correlates to the immunological activity of disease [71].

Membrane insertion of C5b-9, although insufficient to cause cell lysis, does induce cell activation and signal transduction, with increased production of multiple potentially nephritogenic molecules, including oxidants, proteases, cytokines, growth factors, vasoactive molecules, and extracellular matrix. This appears to occur in part through upregulation of glomerular epithelial cells production of transforming growth factor-β (TGF-β) isoforms II and III, as well as increased expression of TGF-β receptors in response to C5b-9. Glomerular injury mediated by C5b-9 induces a nonselective proteinuria through loss of both the size and the charge-selective properties of the glomerular capillary wall. Increased excretion of C5b-9 can be detected in urine.

Characteristic immunofluorescence or immunohistochemistry finding in MN are extensive subepithelial deposits of antibodies and complement components including C3 and C5-9. In 50% of cases, C3 deposits accompany deposits of IgG. They are of the similar diffuse granular pattern and are also localized subepithelially. Positive C3 staining (C3c) reflects active ongoing immune deposit formation and complement activation at the time of the biopsy. Staining for C5b-9 is also present, and C1 and C4 are often absent.

While the nature of the deposited antibody in human MN has not yet been established, many other aspects of the immunopathogenesis of MN are now understood based on studies of the Heymann nephritis models in rats, which closely simulate human lesion.

Tubulointerstitial disease and complement system

Proteinuria is now accepted to be not only a sign of glomerular lesions but also a contributory factor to the development of progressive tubulointerstitial changes. Excellent correlations between the degree of proteinuria and rate of decline of glomerular filtration rate have been demonstrated [72].

The complement system is being increasingly implicated in the pathogenesis of progressive renal disease resulting from persistent proteinuria. Under normal circumstances glomeruli (GBM and layers of endothelial and epithelial cells) restrict protein (including complement proteins) passage into urine and the tubular lumen. This barrier is impaired in GN and tubular cells are exposed to protein-containing urine. Activation of the complement system contributes to tubulointerstitial damage that invariably accompany glomerular diseases [73].

Data are now accumulating that proteins, including complement components derived from leak into the urine in the course of glomerular lesion, cause complement activation on the luminal surface of tubular epithelial cells. Damage of these cells results in fibrotic and inflammatory changes, leading to end-stage renal disease [74].

The complement system has a role in mediating these lesions. Most cells of human body are protected from autologous complement attack by expression of several membrane-bound complement regulatory proteins. The expression of complement regulators is low in the tubular epithelial cells. Tubular epithelial cells express these complement regulatory proteins on their baso-lateral side and not on the luminal side [75, 76].

It has recently been established that renal tubular cells can produce complement proteins, activate complement, and respond to complement activation. Proteins in urine may cause direct activation of the tubular cells to overexpress complement proteins and contribute to local tissue injury [73]. Renal C3 production, mainly at the tubular level, may be induced by urinary excretion of C5-9 in idiopathic membranous glomerulopathy and may have a pathogenetic role in the tubulointerstitial damage. Local synthesis of complement components in the kidney may have a role both in host defense and in the promotion of interstitial inflammation and scaring. The most consistent and dramatic local expression of effectors molecules (e.g., C3) occurs in the proximal tubule. Such tubular complement may be mediator of interstitial damage [9].

Several bacteria have the capacity to colonize the urinary tract and cause infection. Most of these infections do not reach higher than the bladder but some are able to reach kidney and cause pyelonephritis. Various studies indicate that complement plays a role in the defense against bacteria, probably at all levels of infection. However, local production of C3 by tubular epithelial cells is very important in the colonization of the upper urinary tract, since C3-deficient mice exhibited less severe infections compared to C3-sufficient mice [77].

Recent data indicate that complement produced locally is very important in relation to transplant rejection and ischemia-reperfusion injury, since renal transplants from C3-deficient mice into wild-type C3-sufficient mice were not rejected, whereas C3-sufficient kidneys were rejected in this setting [78].

Complement, renal transplant and ischemia/reperfusion injury

Ischemia/reperfusion injury of the kidney is best explained by hypoxia damage of tubular epithelial cells that activate complement by alternative pathway. Tubular epithelial cells produce all the proteins of the alternative pathway. Most of the local injury is due to the assembly of the membrane attack complex [18].

The complement system plays a role in the rejection of xenotransplants [79]. Hyperacute rejection can be attributed to the reactivity of natural antibodies and activation of the complement system. Strategies to prevent or reduce complement activation include complement depletion or inhibition in the recipient or expression of natural complement regulatory proteins, such as decay-accelerating factor and membrane cofactor protein on donor cells, making them resistant to activation of recipient complement [80].

C4d deposits are found in 83% of patients with chronic allograft nephropathy. They are deposited mainly along peritubular capillaries, but they can be localized on glomeruli as well [81]. The exact pathogenetic significance of these deposits is unknown, but they may indicate ongoing humoral immune reaction.

Complement activation in cholesterol crystal emboli disease

Cholesterol crystal emboli (CCE) is still an under-diagnosed condition causing renal dysfunction to the degree of acute renal insufficiency. Clinical observation of low serum complement, peripheral blood eosinophilia and eosinophiluria suggest that activation of complement system participate in inflammatory changes seen on histology specimens in patients with renal CCE [82, 83]. Activation of complement *in vivo* on trapped cholesterol crystals may result in complement cleavage products (C3a and C5a) causing chemotaxis for polymorphonuclear leukocytes and eosinophils with inflammation in small blood vessels.

References

1 Yamamoto T, Wilson CB (1987) Quantitative and qualitative studies of antibody-induced mesangial cell damage in the rat. *Kidney Int* 32: 514–525

2 Quigg RJ (2003) Complement and the kidney. *J Immunol* 171: 3319–3324

3 Trouw LA, Seelen MA, Daha MR (2003) Complement and renal disease. *Mol Immunol* 40: 125–134

4 Wang Y, Hu Q, Madri AA, Rollins SA, Chodera A, Matis LA (1996) Amelioration of lupus-like autoimmune disease in NZB/W F1 mice after treatment with a blocking monoclonal antibody specific for complement component C5. *Proc Natl Acad Sci USA* 93: 8563–8568

5 Thomas TC, Rollins SA, Rother RP, Giannoni MA, Hartman SL, Elliott EA, Nye SH, Matis LA, Squinto SP, Evans MJ (1996) Inhibition of complement activity by humanized anti-C5 antibody and single-chain Fv. *Mol Immunol* 33: 1389–1401

6 Ravirajan CT, Wang Y, Matis LA, Papadaki L, Griffiths MH, Latchman DS, Isenberg DA (2004) Effect of neutralizing antibodies to Il-10 and C5 on the renal damage caused by a pathogenic human anti-dsDNA antibody. *Rheumatology (Oxford)* 43: 442–447

7 Edberg JC, Tosic L, Wright EL, Sutherland WM, Taylor RP (1988) Quantitative analyses of the relationship between C3 consumption, C3b capture, and immune adherence of complement-fixing antibody/DNA immune complexes. *J Immunol* 141: 4258–4265

8 West CD (1989) The complement profile in clinical medicine: Inherited and acquired conditions lowering the serum concentrations of complement component and control proteins. *Complement Inflamm* 6: 49–64

9 Welch TR, Beischel LS, Witte DP (1993) Differential expression of complement C3 and C4 in the human kidney. *J Clin Invest* 92: 1451–1458

10 Wyatt RJ, Forristal J, West CD, Sugimoto S, Curd JG (1988) Complement profiles in acute post-streptococcal glomerulonephritis. *Pediatr Nephrol* 2: 219–223

11 Deppisch R, Ritz E, Hansch GM, Schols M, Rauterberg EW (1994) Bioincompatibility-perspectives in 1993. *Kidney Int* 44 (Suppl): S77–S84

12 Hauser AC, Derefler K, Stockenhuber F, Janata O, Balcke P (1990) Generation of the membrane attack complex during haemodialysis: impact of classical and alternative pathway components. *Clin Sci (Lond)* 79: 471–476

13 Innes A, Farrell AM, Burden RP, Morgan AG, Powell RJ (1994) Complement activation by cellulosic dialysis membranes. *J Clin Pathol* 47: 155–158

14 Lhotta K, Wurzner R, Kroenenberg F, Oppermann M, Konig P (1998) Rapid activation of the complement system by cuprophane depends on complement component C4. *Kidney Int* 53: 1044–1051

15 Brunet P, Jaber K, Berland Y, Baz M (1992) Anaphylactoid reactions during hemodialysis and hemofiltration: role of associating AN69 membrane and angiotensin I- converting enzyme inhibitors. *Am J Kidney Dis* 19: 444–447

16 Varela MP, Kimmel PL, Phillips TM, Mishkin GJ, Lew SQ, Bosch JP (2001) Biocompatibility of hemodialysis membranes: interrelations between plasma complement and cytokine levels. *Blood Purif* 19: 370–379

17 Chenoweth DE (1984) Complement activation during hemodialysis: clinical observations, proposed mechanisms, and theoretical implications. *Artif Organ* 8: 281–290

18 Inal JM, Pascual M, Lesavre P, Schifferli JA (2003) Complement inhibition in renal disease. *Nephrol Dial Transplant* 18: 237–240

19 Kashgarian M, Hayslett JP (1989) Renal involvement in systemic lupus erythematosus. In: C Tischer, B Brenner (eds): *Renal Pathology*. Lippincott, Philadelphia, 389

20 Couser WG (1998) Pathogenesis of glomerular damage in glomerulonephritis. *Nephrol Dial Transplant* 13 (Suppl 1): 10–15

21 Walport MJ (2001) Complement. First of two parts. *N Engl J Med* 344: 1058–1066

22 Walport MJ (2001) Complement. Second of two parts. *N Engl J Med* 344: 1140–1144

23 Pussell BA, Bourke E, Nayef M, Morris S, Peters DK (1980) Complement deficiency and nephritis. A report of family. *Lancet* 1 (8170): 675–677

24 Roord JJ, van Diemen-van, Stennvorde RA, Shuurman HJ, Rijkers GT, Zegers BJ, Gmelig Meyling FH, Stoop JW (1989) Membranoproliferative glomerulonephritis in a patient with congenital dericiency of the third component of complement: effect of treatment with plasma. *Am J Kidney Dis* 205: 595–609

25 Hironaka K, Makino H, Amano T, Ota Z (1993) Immune complex glomerulonephritis in a pregnant woman with congenital C9 deficiency. *Intern Med* 32: 806–809

26 Yoshioka K, Takemura T, Akano N, Okada M, Yagi K, Maki S, Inai S, Akita H, Koitabashi Y, Takekoshi Y (1992) IgA nephropathy in patients with congenital C9 deficiency. *Kidney Int* 42: 1253–1258

27 Rosenberg ME, Girton R, Finkel D, Chmielewski D, Barrie IIIA, Witte DP, Zhu G, Bissler JJ, Harmony JA, Aronow BJ (2002) Apolipoprotein J/clusterin prevents a progressive glomerulopathy of aging. *Mol Cell Biol* 22: 1893–1902

28 Ault BH (2000) Factor H and the pathogenesis of renal diseases. *Pediatr Nephrol* 14: 1045–1053

29 Levy M, Halbach-Mecarelli L, Gubler MC, Kohout G, Bensenouci A, Niaudet P, Hauptmann G, Lesavre P (1986) H deficiency in two brothers with atypical dense intramembranous deposit disease. *Kidney Int* 30: 949–956

30 Hogasen K, Jansen JH (1995) Porcine membranoproliferative glomerulonephritis type II is caused by factor H deficiency. *J Clin Invest* 95: 1054–1061

31 Jansen JH, Hogasen K, Harboe M, Hovig T (1998) *In situ* complement activation in porcine membranoproliferative glomerulonephritis type II. *Kidney Int* 53: 331–349

32 Pickering MC, Cook HT, Warren J, Bygrave AER, Moss J, Walport MJ, Botto M (2002) Uncontrolled C3 activation causes membranoproliferative glomerulonephritis in mice deficient in complement factor H. *Nat Genet* 31:424–428

33 Sadallah S, Gudat F, Laissue JA, Spath PJ, Schifferli JA (1999) Glomerulonephritis in a patient with complement factor I deficiency. *Am J Kidney Dis* 33: 1153–1157

34 Daha MR, Fearon DT, Austen KF (1976) C3 nephritic factor (C3NeF): stabilization of fluid phase and cell-bound alternative pathway convertase. *J Immunol* 116: 1–7

35 Daha MR, van Es LA (1979) Further evidence for the antibody nature of C3 nephritic factor (C3Nef). *J Immunol* 123: 755–758

36 Daha MR, van Es LA (1980) Relative resistance of the F-42 stabilized classical pathway C3 convertase to inactivation by C4-binding protein. *J Immunol* 125: 2051–2054

37 Halbwachs L, Leveille M, Lesavre P, Wattel S, Leibowitch J (1980) Nephritic factor of the classical pathway of complement: immunoglobulin G autoantibody directed against the classical pathway C3 convertase enzyme. *J Clin Invest* 65: 1249–1256

38 Hrick DE, Chung-Park M, Sedor JR (1998) Glomerulonephritis. *N Engl J Med* 339: 888–899

39 Varade WS, Forristal J, West CD (1990) Patterns of complement activation in idiopathic membranoproliferative glomerulonephritis, types I, II and III. *Am J Kidney Dis* 16: 196–206

40 Mathieson PW, Wurzner R, Oliveria DB, Lachmann PJ, Peters DK (1993) Complement-mediated adipocyte lysis by nephritic factor sera. *J Exp Med* 177: 1827–1831

41 Jasin HE (1979) Systemic lupus erythematosus, partial lipodistrophy and hypocomplementemia. *J Rheumatol* 6: 43–50

42 Fremeaux-Bacchi V, Weiss L, Demouchy C, May A, Palomera S, Kazatchkine MD (1995) Hypocomplementaemia of poststreptococcal acute glomerulonephritis is associated with C3 nephritic factor (C3NeF) IgG autoantibody activity. *Nephrol Dial Transplant* 10: 1782–1783

43 Gewurz AT, Imherr SM, Strauss S, Gewurz H, Mold C (1983) C3 nephritic factor and hypocomplementemia in a clinically healthy individual. *Clin Exp Immunol* 54: 253–258

44 Sissons JG, West RJ, Fallows J, Williams DG, Boucher BJ, Amos N, Peters DK (1976) The complement abnormalities of lipodystrophy. *N Engl J Med* 294: 461–465

45 Spitzer RE, Stitzel AE, Tsokos G (1992) On the origin of C3 nephritic factor (antibody

to alternative pathway C3 convertase): evidence for the Adam and Eve concept of autoantibody production. *Clin Immunol Immunopathol* 64: 177–183

46 Mollnes TE, Ng YC, Peters DK, Lea T, Tshopp J, Harboe M (1986) Effect of nephritic factor on C3 and on the terminal pathway of complement *in vivo* and *in vitro*. *Clin Exp Immunol* 65: 73–79

47 Tanuma Y, Ohi H, Hatano M (1990) Two types of C3 nephritic factor: properdin-dependent C3NeF and properdin-independent C3NeF. *Clin Immunol Immunopathol* 56: 226–238

48 Ohi H, Yasugi T (1994) Occurrence of C3 nephritic factor and C4 nephritic factor in membranoproliferative glomerulonephritis (MPGN). *Clin Exp Immunol* 95: 316–321

49 Jokiranta TS, Solomon A, Pangburn MK, Zipfel PF, Meri S (1999) Nephritogenic lambda light chain dimmer: a unique human miniautoantibody against complement factor H. *J Immunol* 163: 4590–4596

50 Horvath L, Czirjak L, Fekete B, Jakab L, Pozsonyi T, Kalabay L, Romics MR, Miklos K, Varga L, Prohaszka Z et al (2001) High levels of antibodies against C1q are associated with disease actibvity and nephritis but not with other organ manifestations in SLE patients. *Clin Exp Rheumatol* 19: 667–672

51 Siegert C, Daha M, Westedt ML, van der Voort E, Breedveld F (1991) IgG autoantibodies against C1q are correlated with nephritis, hypocomplementemia and dsDNA antibodies in systemic lupus erythematosus. *J Rheumatol* 18: 230–234

52 Siegert CE, Daha MR, Halma C, van der Voort EA, Breedveld FC (1992) IgG and IgA autoantibodies to C1q in systemic and renal diseases. *Clin Exp Rheumatol* 10: 19–23

53 Coremans IE, Spronk PE, Bootsma H, DSaha MR, van der Voort EA, Kater L, Breedveld FC, Kallenberg CG (1995) Changes in antibodies to C1q predict renal relapses in systemic lupus erythematosus. *Am J Kidney Dis* 26: 595–601

54 Trendelenburg M, Marfurt J, Gerber I, Tyndall A, Schifferli JA (1999) Lack of occurrence of severe lupus nephritis among anti-C1q autoantibody negative patients. *Arthritis Rheumatol* 42: 187–188

55 Mannik M, Werner MH (1997) Deposition of antibodies to the collagen-like region of C1q in renal glomeruli of patients with proliferative lupus glomerulonephritis. *Arthritis Rheum* 40: 1504–1511

56 Siegert CE, Daha MR, Swaak AJ, van der Voort EA, Breedveld FC (1993) The relationship between serum titers of autoantibodies to lupus erythematosus. *Clin Immunol Immunopathol* 67: 204–209

57 Pickering MC, Botto M, Taylor PR, Lachmann PJ, Walport MJ (2000) Systemic lupus erythematosus, complement deficiency, and apoptosis. *Adv Immunol* 76: 227–324

58 Walport MJ, Davies KA, Morley BJ, Botto M (1997) Complement deficiency and autoimmunity. *Ann NY Acad Sci* 815: 267–281

59 Lachmann PJ (1961) An attempt to characterize the lupus erythematosus cell antigen. *Immunology* 4: 153–163

60 Pan CG, Strife CF, Ward MK, Spitzer RE, McAdams AJ (1992) Long-term follow-up of

non-systemic lupus erythematosus glomerulonephritis in patients with hereditary angioedema: report of four cases. *Am J Kidney Dis* 19: 526–531

61 Wisnieski JJ, Emancipator SN, Korman NJ, Lass JH, Zaim TM, McFadden ER (1994) Hypocomplementemic urticarial vasculitis syndrome in identical twins. *Arthritis Rheum* 37: 1105–1111

62 Emancipator SN (1994) IgA nephropathy: morphologic expression and pathogenesis. *Am J Kidney Dis* 23: 451–462

63 O'Donoghue DJ, Darvill A, Baillardie W (1991) Mesangial cell autoantigens in immunoglobulin A nephropathy and henoch-Schonlein Purpura. *J Clin Invest* 88: 1522–1530

64 Rits M, Hiemstra PS, Bazin H, Van Es LA, Voerman JP, Daha MR (1988) Activation of rat complement by soluble and insoluble rat IgA immune complexes. *Eur J Immunol* 18: 1873–1880

65 Rauterberg EW, Lieberknecht HM, Wingen AM, Ritz E (1987) Complement membrane attack (MAC) in idiopathic IgA-glomerulonephritis. *Kidney Int* 31: 820–829

66 Brandt J, Pippin J, Shulze M, Mansche GM, Alpers CE, Johnson RJ, Gordon K, Couser WG (1996) Role of the complement membrane attack complex (C5-9) in mediating experimental mesangioproliferative glomerulonephritis. *Kidney Int* 49: 335–343

67 Nangaku M, Pippin J, Shankland SJ, Johnson RJ, Couser WG (1997) Renal microvascular injury induced by the antibody to glomerular endothelial cells is mediated by C5b-9. *Kidney Int* 52: 1570–1578

68 Couser WG (1999) Glomerulonephritis. *Lancet* 353: 1509–1515

69 Cameron JS, Vick RM, Ogg CS, Seymour WM, Chantler C, Turner DR (1973) Plasma C3 and C4 concentrations in management of glomerulonephritis. *Brit Med J* 3: 668–672

70 Floege J, Feehally J (2000) IgA nephropathy: recent developments. *J Am Soc Nephrol* 11: 2395–2403

71 Montinaro V, Lopez A, Monno R, Cappiello V, Manno C, Gesualdo L, Schena FP (2000) Renal C3 synthesis in idiopathic membranous nephropathy: correlation to urinary C5b-9 excretion. *Kidney Int* 57: 137–146

72 Newman DJ, Thakkar H, Gallagher H (2000) Progressive renal disease: does the quality of the proteinuria matter or only the quantity? *Clin Chim Acta* 297: 43–54

73 Tang S, Lai KN, Sacks SH (2002) Role of complement in tubulointerstitial injury from proteinuria. *Kidney Blood Press Res* 25: 120–126

74 Van Kooten C, Langers AM, Bruijn JA, Daha MR (1999) Role of tubular cells in progressive renal disease. *Kidney Blood Press Res* 22: 53–61

75 Ischida S, Yuzawa Y, Okada H, Yoshioka K, Matsuo S (1994) Localization of the complement regulatory proteins in the normal human kidney. *Kidney Int* 46: 89–96

76 Nangaku M (1998) Complement regulatory proteins in glomerular diseases. *Kidney Int* 54: 1419–1428

77 Springall T, Sheerin NS, Abe K, Holers VM, Wan H, Sacks SH (2001) Epithelial secretion of C3 promotes colonization of the upper urinary tract by *Escherichia coli*. *Nat Med* 7: 801–806

78 Pratt JR, Basheer SA, Sacks SH (2002) Local synthesis of complement component C3 regulates acute renal transplant rejection. *Nat Med* 8: 582–587

79 Dooldeniya MD, Warrens AN (2003) Xenotransplantation: where are we today? *J R Soc Med* 96: 111

80 Coopper DK, Gollackner B, Sach DH (2002) Will pig solve the transplantation backlog? *Annu Rev Med* 53: 133

81 Mauiyyedi S, Pelle PD, Saidman S, Collins AB, Pascual M, Tolkoff-Rubin NE, Williams WW, Cosimi AA, Schneeberger EE, Colvin RB (2001) Chronic humoral rejection: identification of antibody-mediated chronic renal allograft rejection by C4d deposits in peritubular capillaries. *J Am Soc Nephrol* 12: 574–582

82 Wilson DM, Salazer TL, Farkouh ME (1991) Eosinophiluria in atheroembolic renal disease. *Am J Med* 91: 186–189

83 Cosio FG, Zager RA, Sharma HM (1985) Atheroembolic renal disease causes hypocomplementemia. *Lancet* 2: 118–121

Complement in renal transplantation

Wuding Zhou and Steven H. Sacks

Department of Nephrology and Transplantation, Guy's, King's and St. Thomas' School of Medicine, King's College of London, London, SE1 9RT, UK

Introduction

Over the past two decades, renal transplantation has become a highly successful treatment for end-stage kidney disease, with 85% of kidney grafts still functioning at the end of the first year. Nonetheless, a significant number of grafts undergo damage in the early period, and the incidence of late graft loss due to chronic rejection has not substantially changed [1]. It is thought that both alloantigen-dependent and -independent mechanisms contribute to this late graft loss. It is, therefore, vital that all potential mechanisms of graft injury are considered, especially where the mechanism of injury may not be controlled by current immunosuppressive drug regimens, such as injuries caused by intrinsic stimuli including complement activation in the kidney, ischemia/reperfusion (I/R) and surgical procedure. In this chapter, we discuss the complement system in renal transplantation and potential contributions of locally produced complement in the kidney to renal graft injury or rejection.

Complement and renal transplantation

The complement system is pivotal in the regulation of inflammation and host defense (reviewed in [2]). The complement system consists of a set of distinct plasma proteins that react in a cascade manner to opsonize pathogens, and induce a series of inflammatory responses. It can be activated through three pathways, the classical pathway, the mannose-binding lectin (MBL) pathway and the alternative pathway, which are triggered by a number of stimuli including bound antibodies, pathogen surfaces and spontaneously hydrolyzed C3. The function of the complement system in the regulation of inflammation and host defense is achieved by complement split products (e.g., C3b, C4b, C3a and C5a) in cooperation with antibody and phagocytes, as well as the formation complement membrane attack complex (MAC). Complement activation is well controlled by numerous complement inhibitors and regulators under normal circumstances. However, in the event of

Complement and Kidney Disease, edited by Peter F. Zipfel

inappropriate or excessive activation, complement can cause tissue injury. In addition to their role in host defense and stimulation of a nonspecific inflammatory response, complement also participates in the regulation of the antigen-specific immune response including T cells and B cells (reviewed in [2, 3]).

Renal transplantation is an intricate procedure; the organ graft could suffer I/R injury, hyperacute rejection, acute rejection and chronic rejection as well as infection. All of these events are associated with the development of nonspecific inflammatory injury and antigen-specific immune response in the organ graft and recipient. Given the functions in both the innate and adaptive immune responses, complement may play an important role in renal transplantation.

Role of complement in renal I/R injury

I/R injury is an important form of injury that occurs upon reperfusion of vascularized tissue after an extended period of ischemia. It is an unavoidable event in organ transplantation. Numerous clinical and experimental studies have shown that renal I/R injury has a major impact on short- and long-term graft survival after organ transplantation [4–6]. During I/R insult, the depletion of ATP causes intracellular accumulation of ions and water, resulting in cell swelling. Subsequently, multiple enzyme systems associated with the metabolism of oxygen, lipid, phospholipid, nucleotide are activated, leading to cytoskeleton disruption, membrane damage, DNA degradation and eventually cell death. This injury becomes manifest through the participation of a number of components such as complement activation, molecular oxygen, neutrophils and adhesion molecules such as P-selectin [7–9]. More recently, T cells and B cells have also been considered as components contributing to the process of reperfusion injury [10–12].

Complement activation is an early event in the course of reperfusion injury, which is evident by detecting activated complement product on renal tubules as early as 30 min after reperfusion [13]. The generation of complement effector molecules may influence the function of other factors, such as free radicals, neutrophils and the products of activated endothelium [14]. Thus, activation of complement is an important event in the setting of I/R injury. Complement activation after hypoxia and reperfusion causes vascular and parenchymal cell injury, which may be mediated through a variety of effector products. Complement activation releases a number of biologically active products, several of which possess pro-inflammatory activity *in vitro*. The early products C4a, C3a and C5a, the anaphylatoxins, can induce smooth muscle contraction, cause the release of histamine, and lead to increased vascular permeability [15]. In addition, C5a can act directly on neutrophils, promoting chemotaxis and activation, and can act on both neutrophils and endothelium to up-regulate cell adhesion molecules such as CD11b/CD18 and intercellular adhesion molecule (ICAM-1) [15, 16]. The MAC, C5b-9, inserts into the membrane of target cells, directly causing cell injury and necrosis [17, 18]. Sub-lethal amounts of C5b-9 can activate neutrophils as well as endothelium and epithelium by up-regulating adhesion molecules, and promoting the release of cell stimulants such as hydrolytic enzymes, reactive oxygen species, arachidonic acid metabolites and cytokines [19–22]. In addition, C5b-9 can enhance the pro-coagulant properties of endothelium [23].

Although the role of complement in I/R injury has been widely studied in a number of organs such as heart, lung, brain, intestine and muscle [24–27], less was known in renal I/R injury until recently. Using various complement-deficient mice, we demonstrated that complement plays an important pathogenic role in renal I/R injury [28]. Our study showed that renal injury is reduced up to 50% in C3-defcient mice compared with their wild-type counterparts. Because a similar degree of protection against renal I/R injury is achieved in C3- and C5- or C6-deficient mice, this suggests that products derived from the terminal pathway of complement activation is an important factor in pathogenesis of renal I/R injury. Furthermore, C4 was found to be dispensable for the development of renal I/R injury, indicating that complement activation in the mouse model occurs through the alternative pathway, and is less or not dependent on the classical or lectin pathways. Subsequently, renal I/R injury has been studied in mice with deficiency in Factor B, an essential component of the alternative pathway [29]. These studies found that Factor B-deficient mice develop substantially less renal I/R injury, confirming that complement activation during renal I/R injury occurs mainly via the alternative pathway. Renal I/R injury has also been studied in mice with deficiency in complement inhibitors CD55 (decay accelerating factor, DAF) and CD59 [30, 31]. CD55 inhibits complement activation by promoting the breakdown of activated C3 and accelerating decay of the C3 converting enzyme complexes; CD59 protects against complement-mediated cell injury by blocking the formation of the MAC. These two studies showed that mice lacking either CD55 or CD59 are highly susceptible to renal I/R injury [30, 31]. Moreover, double deficiency of CD55/CD59 significantly exacerbates the injury [30]. Thus, complement regulatory proteins may be equally important as complement components in renal I/R injury. This is supported by the previous observations that complement regulatory proteins have reduced or diminished expression in I/R-injured tissue and in hypoxic cells [32, 33]. A more recent study using P-selectin inhibition in C3-deficient mice found that treatment with anti-P-selectin antibody to reduce the neutrophil-mediated renal I/R injury was equally effective in the absence of C3, suggesting that complement- and P-selectin-mediated pathways of renal I/R injury are mutually independent [13]. Thus, complement and P-selectin may pose distinct targets for therapy.

Besides investigation in complement-deficient animals, renal I/R injury has also been studied using complement inhibitory reagents, which include a membrane-targeted portion of the complement regulator CR1, C5aR antagonist, anti-C5 mAb and Crry-Ig. Perfusion and incubation of donor kidney with membrane-targeted complement regulator CR1 significantly protected transplanted tissue from I/R injury, and improved isograft function in the short- and long-term after transplantation [34]. C5a blockade with C5a receptor antagonist prevented many of the features of renal injury in rats and mice, suggesting that C5a is an important pathogenic product in renal I/R injury, and that C5aR antagonist may be a useful therapeutic reagent for preventing such injury. The results with C5aR antagonists in mice also suggest that involvement of C5a in the pathogenesis of renal I/R injury is both neutrophil dependent and neutrophil independent, indicating that C5a receptor on the target organ may also be an important mediator of I/R injury [35, 36]. Treatment with anti-C5-mAb, which inhibits the generation of C5a and the formation of C5b-9, prevented C5b-9 formation on the tubule epithelium and coordinately reduced renal I/R injury [37]. Treatment with Crry-Ig, which inhibits the enzymes activating C3 and C5, did not show significant reduction of renal I/R injury, although reduction of renal I/R injury was observed in $C3^{-/-}$ mice in the same study [38, 39]. The reason for this discrepancy is not clear, but may reflect the importance of successful tissue penetration of therapeutics in renal I/R injury.

In general, utilizing complement inhibitory reagents in renal I/R injury seems promising. However, to achieve more effective therapeutic complement inhibition in the kidney, further studies are needed to better understand the precise mechanism of renal I/R injury. Because the precise mechanism of I/R injury may vary from one organ to another, this may give a different slant regarding the approach to therapy. The classical paradigm for complement-mediated reperfusion injury is seen in the heart, gut and muscle. In these organs, post-ischemic injury initially may cause small vessel thrombosis, vessel wall inflammatory cell infiltration and direct endothelial membrane injury. Tissue infarction appears to be downstream event following blood vessel damage. Ig deposition on damaged vasculature may favor classical pathway activation, with secondary amplification by the alternative pathway loop. In contrast to theses organs, renal post-ischemic injury of mild and moderate severity appears to involve primary damage of the renal tubules, those worst affected being in the hypoxia-sensitive region of the corticomedullary junction [7, 28]. As mentioned earlier, MAC appears to cause a significant amount of tubule damage. C5a may have a direct action on the renal tubule, in addition to an effect on infiltrating neutrophils [36]. Therefore, reagents with good tissue penetration that are able access the tubulointerstitial space, and reagents only targeting the alternative pathway, or more specifically, the terminal pathway, should be considered in the design of a therapeutic strategy for the prevention of renal I/R injury.

Complement in hyperacute rejection

Currently, a major difficulty facing the transplant community is the shortage of donor organs. A promising solution at the moment is xenotransplantation, using organs from species discordant to man, such as pig kidney (reviewed in [40]). However, the earliest and most striking immunological obstacle in xenotransplantation is hyperacute rejection (reviewed in [41, 42]). Similarly, in pre-sensitized recipients, hyperacute allograft rejection is an important potential source of graft loss [43, 44]. Rapid rejection of an organ graft may occur within several minutes of revascularization, and is characterized pathologically by vascular thrombi and congestion, vascular endothelial damage, interstitial hemorrhage, edema and infarction (reviewed in [45, 46]) Complement activation is a critical mediator of hyperacute rejection [47–49]. It is thought that complement activation is mainly triggered by pre-existing antibody through the classical pathway [50], although the alternative pathway may also be involved in the initiation of complement activation [51–53]. Products of complement activation, such as MAC, C3a and C5a could contribute to hyperacute rejection directly and/or indirectly [54, 55].

In addition, complement regulatory proteins play a critical role in determining the intensity of hyperacute rejection. These regulatory proteins include CD55, CD46 (membrane co-factor protein, MCP) and CD59, which control complement activation at different levels of the complement cascade. However, these regulatory proteins often show homologous restriction, which means they may only regulate complement activation in their own species. With great advances in molecular biology, a number of transgenic pigs expressing human DAF (hDAF), human MCP (hCD46) or human CD59 (hCD59) have been generated [56–60]. Several studies, transplanting donor kidney from complement regulator-transgenic pig to baboon, have shown that inhibition of complement activation on donor organs effectively reduced hyperacute rejection. The xenografts, by transplanting kidney from hDAF-pig to baboon, with different protocols of immunosuppression, survived between 1 and 31 days [61]. Baboons, receiving kidneys from double transgenic pigs expressing hCD55 and hCD59 without immunosuppression, survived for 5 and 6 days (versus baboon transplanted with non-transgenic kidneys surviving 2 days) [62]. In transplantion of hCD46 transgenic pig kidney into adult baboon, where recipients did not receive immunosuppression, the xenografts showed significantly reduced macroscopic damage compared with normal pig renal graft [63]. These studies demonstrated that the expression of complement regulator on pig organs provides some protection against complement-mediated hyperacute rejection. This protection, however, is short-lived, the longest median time for organ survival being under a month. Many other obstacles remain in xenotransplantation, such as acute humoral rejection and the potential risk of transmission of animal infections.

Complement in allograft rejection

Although it is well established that complement activation plays a significant role in mediating renal I/R injury and hyperacute rejection, less is known about the role of the complement system in unsensitized allograft rejection. Recent studies have drawn attention to the role of complement in acute allograft rejection. Analysis of human tissue biopsy sepecimens provides strong evidence of complement activation following kidney transplantation. The demonstration of C3 split product in the tubular basement membrane and C4 split product in capillaries of allografts undergoing early graft dysfunction suggests an association between complement activation and graft outcome [64, 65].

Animal models of complement deficiency and complement inhibition have been widely used in transplantation studies. These studies showed that recipients with a deficiency in different complement components or those treated with complement inhibitors demonstrate prolonged survival of skin, heart, lung, as well as kidney allografts [54, 66–70]. Early studies in rat renal allografts (a fully MHC-mismatched strain combination, Lewis to DA), using the soluble inhibitor complement receptor-1 to treat recipient rats but with no other treatment, have found that inhibition of complement has a beneficial effect on graft outcomes. These studies showed that complement inhibition significantly improves the outcome of renal allografts, prolongs renal allograft survival and reduces the anti-donor T cell response. In addition, the deposition of C3 and MAC on tubules and vessels is almost completely abolished in complement-inhibited rats [69]. Furthermore, complement-inhibited rats display marked impairment of antibody production against donor antigen [70]. In addition, a study in a mouse skin allograft model by Marsh et al. [67] showed that mice with complement deficiency either of C3 or C4, have a significantly reduced titers of alloantibody and a failure of Ig class switching to high-affinity IgG. More recently, another study using a membrane-targeted complement regulator derived from CR1, which has been shown to be effective in preventing renal I/R injury as mentioned above, also showed that rat renal allograft rejection was significantly reduced [34].

Together, these studies provide support that complement activation is associated with renal allograft rejection. Complement inactivation results in prolonged graft survival, reduced nonspecific inflammatory injury of the organ graft and impaired T cell and B cell responses against donor antigen.

Locally synthesized C3 in allograft rejection

Increasing recognition that extrahepatic tissue may be a source of complement raises the question of whether kidney is able to produce complement components, and if local production participates in the pathogenesis of renal injury. Since the 1980s, when Colten and colleagues [71] first detected complement synthesis in murine kidney, and Feucht et al. [72] subsequently detected C4 mRNA in normal human kidney tissue, many studies have been carried out to investigate intrarenal synthesis of complement. This subject has been extensively reviewed elsewhere [73]. In brief, renal tissue has the capacity to synthesize most components of the classical and alternative activation pathways of complement. At least four types of resident kidney cells (mesangial, endothelial, glomerular and tubular epithelial) are capable of synthesizing multiple complement proteins either spontaneously or in response to cytokine stimulation. It thus appears that extravascular production of complement is sufficient for activation of the complement cascade.

Gene expression studies on human kidney biopsy material revealed that expression of C3 and C4 mRNA is markedly up-regulated in a variety of inflammatory conditions [74–76]. This is especially so in rejecting kidney allografts [76]. Using an ELISA that is specific for the donor C3, Tang et al. [77] found that a single donor kidney contributes 4–5% of the total circulating C3, which increases during allograft rejection. Using a donor-specific antibody staining method, Andrews at al. [78] found that renal tubular epithelium is likely to be the most abundant source of C3 in inflamed human kidney. These studies together suggest a strong linkage between the local synthesis of C3 in human kidney grafts and graft rejection.

However, the functional significance of local production of C3 needs to be clarified. The emergence of complement-deficient mice and the improvement of microsurgery technique for mouse kidney transplantation in the last decade, allowed us to clarify the functional significance of local production of C3 in allograft rejection [79, 80]. Our studies found that C3-deficient mouse kidneys (C57BL/6, H-2b), when transplanted into allogeneic wild-type mice (B10Br, H-2k), mostly survived for at least 100 days, compared to only 14 days for C3-sufficient donor mouse kidney [81]. This suggests that locally produced C3 plays an important role in transplant rejection. Specifically, the results suggested that C3-producing cells in the donor kidney had a significant effect on up-regulating the alloantigen-specific immune response.

Complement regulates the immune response in allograft rejection

How complement, particularly the locally synthesized C3 produced by donor kidney, affects allograft rejection is unclear. There are several possible explanations. Firstly, local production of C3 might enhance the complement-mediated nonspecific inflammatory response, such as direct injury on endothelium and epithelium, the infiltration of inflammatory cells and the secretion of pro-inflammatory cytokines, oxygen free radicals and vasoactive substrates from injured cells [15–21]. Consequently, this nonspecific inflammatory response might enhance the organ graft immunogenicity [82]. That is, local inflammation mediated by complement activation could increase the capacity of antigen-presenting cells (APCs) to stimulate alloreactive T cells. This could involve an increase in the number of activated APCs migrating from donor kidney to the recipient secondary lymphoid tissues, and/or an increase in the expression of MHC and co-stimulatory molecules by APCs, including donor and recipient APCs.

Secondly, local production of C3 might have a direct effect on the T cell response. Complement effector products, such C3a and C5a, can bind to complement receptors on T cells, which leads to T cell activation [83, 84]. C3 split products (C3b or C3d), either deposited on APCs or bound to antigen, can potentially enhance antigen uptake, processing and presentation [85–87]. In addition, deposition of C3 split product on tubular epithelial cells can significantly enhance the stimulation of alloreactive T cells [88].

Thirdly, based on observations that C3-positive APCs including dendritic cells (DCs), macrophages and epithelial cells stimulate alloreactive T cells more vigorously than C3-negative APCs ([81] and our unpublished data), it is possible that endogenous C3 produced by donor APCs might have an effect on antigen presentation by these cells and enhance the potency of APCs to stimulate alloreactive T cells.

In addition to an effect on the T cell response, complement also regulates the B cell response in allograft rejection [67, 70]. The mechanism of complement-mediated stimulation of the B cell response appears to involve at least two processes. Co-ligation of B cell antigen receptor and co-receptor (complement receptor 2, CR2) by C3d-coated antigen results in an enhanced signal for B cell activation, which markedly lowers the threshold for the B cell response [89]. Secondly, C3d-coated antigen is more effectively retained in the germinal centers of secondary lymphoid tissues, where rapid proliferation and differentiation of B cells occur. This enhanced trapping of opsonized antigen appears to be the result of binding with complement receptor (CR1/2) expressed on follicular DCs in the germinal centers (reviewed in [90]).

C4d in allograft rejection

Recently, the Banff classification of allograft rejection has been revised by incorporating the presence of deposits of C4d in the interstitial capillaries (with or without the presence of circulating anti-donor antibodies) into the definition of acute humoral rejection [91]. C4 is cleaved during classical pathway activation, and forms a covalent bond with the cell membrane close to the site of antibody attachment. Bound C4b is rapidly degraded by proteolytic cleavage, leaving C4d stably attached to the cell membrane.

Capillary deposition of C4d in renal grafts was first described in 1993 by Feucht and colleagues [65]. They found that in the setting of delayed graft function, especially when recipients were pre-sensitized by donor antigen, peritubular capillary C4d is present in 50% of biopsy specimens. More recently, in a growing number of transplant studies, immunohistological detection of C4d in the interstitial capillaries has become recognized as a sensitive index of antibody-mediated alloreaction [92–97].

The largest of these studies is from Basle [94]. In a cohort of 398 renal transplant biopsies, peritubular capillary C4d was present in 30% of samples. Its presence was indicative of acute humoral rejection, and correlated with a high frequency of panel-reactive antibodies in the recipient serum. Other studies have reported similar findings, and, in addition, have shown an adverse effect of C4d-positive staining on graft outcome [95, 96]. A group in Boston has found that the specificity and sensitivity of peritubular capillary C4d for correctly identifying recipients with anti-donor antibodies was greater than 90% [97]. Others have reported a lower frequency (about 50%) of C4d-positive staining, and it is clear that there is a need for prospective data collection, ideally on unselected cases [92]. In addition to the presence of C4d in grafts with acute humoral rejection, a study of 213 late renal transplant biopsy specimens, from Vienna, showed that the presence of peritubular capillary C4d (in 34%) is associated with chronic transplant glomerulopathy and cellular infiltration [98]. Consequently, C4d detection looks set to become a new tool with which to re-evaluate the role of antibodies in chronic rejection [93].

Clinical implications

Although studies examining the clinical relevance of these observations are in their infancy, several general observations can be made. First, complement activation does occur in human renal allografts, although this may not have been so well appreciated before the application of immunohistological reagents specific for stable tissue-bound activation products, such as the metabolites C3d and C4d [99]. Second, as mentioned above, donor kidney production of complement may be so apparent in the inflamed human transplant that changes are easily detected in the recipient circulation [76, 77]. Third, an earlier suggestion that the donor kidney expression of C3 may have an influence on graft outcome [100] has recently been confirmed in a large cohort of transplant recipients where the influence of the donor C3 allotypes has been again studied (our unpublished results). Fourth, the involvement of the complement cascade may serve as a marker for previously unrecognized humoral rejection in chronic as well as acute graft damage, as inferred above. Fifth, with the growing interest in ABO incompatible transplants, as a measure to overcome the donor organ shortage, protection from complement-mediated endothelial damage can be seen as a growing priority [101].

These observations make a persuasive argument for the development of therapeutic complement inhibitors and their application in renal transplantation. It is not certain that existing agents used to prevent graft rejection have an effect on complement-mediated injury. *In vitro* studies suggest that calcineurin inhibitors, in therapeutic doses, have no effect on local synthesis of complement [102]. Moreover, the effect of prednisolone on complement production and complement-mediated inflammation may be variable [103–105]. Therefore, specific complement inhibition is likely to be needed. It is beyond the scope of this chapter to discuss in detail the approaches currently being undertaken. In general terms, therapeutic complement inhibitors fall into two main categories: soluble and membrane targeted. Soluble complement inhibitors/regulators are likely to be more effective at blocking the effects of circulating complement, for example in antibody-mediated endothelial inflammation. Tissue-targeted inhibitors, on the other hand, may be more appropriate for preventing injury in the extracapillary space. One such approach used extracorporeal treatment of donor kidney with membrane-targeted complement regulator [34]. This avoided systemic complement depletion, yet provided effective reduction of injury of the renal tubule inflammation subsequent to transplantation. A further consideration is whether to target the early or late part of the complement cascade. Prevention of the immunostimulatory functions of complement would require inhibition at the level of C3, whereas terminal cascade inhibition would provide selective avoidance of membrane injury.

Conclusions

Complement poses a relatively new target for the prevention of renal injury in solid organ transplantation. However, the feasibility and efficacy of such treatment remains to be assessed in transplant patients. Based on observation in animal studies, complement inhibition, at the very least, may be expected to lessen I/R injury of the transplanted kidney, including downstream effects on acute and late graft loss. A potential attraction is that by reducing the proinflammatory and immunoregulatory effects of complement, the development of donor-specific immunity might be reduced. Fuller understanding of the mechanisms and sites of action of complement in solid organ transplantation will enable a more logical approach regarding the choice, timing and method of delivery of therapeutic inhibitors.

References

1 Cecka JM (2001) The UNOS renal transplant registry. *Clin Transpl* 1–18

2 Walport MJ (2001) Complement. First of two parts. *N Engl J Med* 344: 1058–1066

3 Walport MJ (2001) Complement. Second of two parts. *N Engl J Med* 344: 1140–1144

4 Troppmann C, Gillingham KJ, Benedetti E, Almond PS, Gruessner RW, Najarian JS, Matas AJ (1995) Delayed graft function, acute rejection, and outcome after cadaver renal transplantation. The multivariate analysis. *Transplantation* 59: 962–968

5 Pagtalunan ME, Olson JL, Tilney NL, Meyer TW (1999) Late consequences of acute ischemic injury to a solitary kidney. *J Am Soc Nephrol* 10: 366–373

6 Herrero-Fresneda I, Torras J, Cruzado JM, Condom E, Vidal A, Riera M, Lloberas N, Alsina J, Grinyo JM (2003) Do alloreactivity and prolonged cold ischemia cause different elementary lesions in chronic allograft nephropathy? *Am J Pathol* 162: 127–137

7 Bonventre JV (1993) Mechanisms of ischemic acute renal failure. *Kidney Int* 43: 1160–1178

8 Sutton TA, Molitoris BA (1998) Mechanisms of cellular injury in ischemic acute renal failure. *Semin Nephrol* 18: 490–497

9 Padanilam BJ (2003) Cell death induced by acute renal injury: a perspective on the contributions of apoptosis and necrosis. *Am J Physiol Renal Physiol* 284: F608–F627

10 Rabb H, Daniels F, O'Donnell M, Haq M, Saba SR, Keane W, Tang WW (2000) Pathophysiological role of T lymphocytes in renal ischemia-reperfusion injury in mice. *Am J Physiol Renal Physiol* 279: F525–F531

11 Burne MJ, Daniels F, El Ghandour A, Mauiyyedi S, Colvin RB, O'Donnell MP, Rabb H (2001) Identification of the CD4(+) T cell as a major pathogenic factor in ischemic acute renal failure. *J Clin Invest* 108: 1283–1290

12 Burne-Taney MJ, Ascon DB, Daniels F, Racusen L, Baldwin W, Rabb H (2003) B cell deficiency confers protection from renal ischemia reperfusion injury. *J Immunol* 171: 3210–3215

13 Farrar CA, Wang Y, Sacks SH, Zhou W (2004) Independent pathways of p-selectin and complement-mediated renal ischemia/reperfusion injury. *Am J Pathol* 164: 133–141

14 Kilgore KS, Friedrichs GS, Homeister JW, Lucchesi BR (1994) The complement system in myocardial ischaemia/reperfusion injury. *Cardiovasc Res* 28: 437–444

15 Hugli TE (1986) Biochemistry and biology of anaphylatoxins. *Complement* 3: 111–127

16 Foreman KE, Glovsky MM, Warner RL, Horvath SJ, Ward PA (1996) Comparative effect of C3a and C5a on adhesion molecule expression on neutrophils and endothelial cells. *Inflammation* 20: 1–9

17 Biancone L, David S, Della P, V, Montrucchio G, Cambi V, Camussi G (1994) Alternative pathway activation of complement by cultured human proximal tubular epithelial cells. *Kidney Int* 45: 451–460

18 David S, Biancone L, Caserta C, Bussolati B, Cambi V, Camussi G (1997) Alternative pathway complement activation induces proinflammatory activity in human proximal tubular epithelial cells. *Nephrol Dial Transplant* 12: 51–56

19 Kilgore KS, Schmid E, Shanley TP, Flory CM, Maheswari V, Tramontini NL, Cohen H, Ward PA, Friedl HP, Warren JS (1997) Sublytic concentrations of the membrane attack complex of complement induce endothelial interleukin-8 and monocyte chemoattractant protein-1 through nuclear factor-kappa B activation. *Am J Pathol* 150: 2019–2031

20 Hansch GM, Seitz M, Betz M (1987) Effect of the late complement components C5b-9 on human monocytes: release of prostanoids, oxygen radicals and of a factor inducing cell proliferation. *Int Arch Allergy Appl Immunol* 82: 317–320

21 Seeger W, Suttorp N, Hellwig A, Bhakdi S (1986) Noncytolytic terminal complement complexes may serve as calcium gates to elicit leukotriene B4 generation in human polymorphonuclear leukocytes. *J Immunol* 137: 1286–1293

22 Campbell AK, Morgan BP (1985) Monoclonal antibodies demonstrate protection of polymorphonuclear leukocytes against complement attack. *Nature* 317: 164–166

23 Wiedmer T, Esmon CT, Sims PJ (1986) Complement proteins C5b-9 stimulate procoagulant activity through platelet prothrombinase. *Blood* 68: 875–880

24 Weisman HF, Bartow T, Leppo MK, Marsh HC Jr, Carson GR, Concino MF, Boyle MP, Roux KH, Weisfeldt ML, Fearon DT (1990) Soluble human complement receptor type 1: *in vivo* inhibitor of complement suppressing post-ischemic myocardial inflammation and necrosis. *Science* 249: 146–151

25 Eppinger MJ, Deeb GM, Bolling SF, Ward PA (1997) Mediators of ischemia-reperfusion injury of rat lung. *Am J Pathol* 150: 1773–1784

26 Huang J, Kim LJ, Mealey R, Marsh HC, Jr., Zhang Y, Tenner AJ, Connolly ES Jr, Pinsky DJ (1999) Neuronal protection in stroke by an sLex-glycosylated complement inhibitory protein. *Science* 285: 595–599

27 Ikai M, Itoh M, Joh T, Yokoyama Y, Okada N, Okada H (1996) Complement plays an essential role in shock following intestinal ischaemia in rats. *Clin Exp Immunol* 106: 156–159

28 Zhou W, Farrar CA, Abe K, Pratt JR, Marsh JE, Wang Y, Stahl GL, Sacks SH (2000) Predominant role for C5b-9 in renal ischemia/reperfusion injury. *J Clin Invest* 105: 1363–1371

29 Thurman JM, Ljubanovic D, Edelstein CL, Gilkeson GS, Holers VM (2003) Lack of a functional alternative complement pathway ameliorates ischemic acute renal failure in mice. *J Immunol* 170: 1517–1523

30 Yamada K, Miwa T, Liu J, Nangaku M, Song WC (2004) Critical protection from renal ischemia reperfusion injury by CD55 and CD59. *J Immunol* 172: 3869–3875

31 Turnberg D, Botto M, Lewis M, Zhou W, Sacks SH, Morgan BP, Walport MJ, Cook HT (2004) CD59a deficiency exacerbates ischemia-reperfusion injury in mice. *Am J Pathol* 165: 825–832

32 Vakeva A, Morgan BP, Tikkanen I, Helin K, Laurila P, Meri S (1994) Time course of complement activation and inhibitor expression after ischemic injury of rat myocardium. *Am J Pathol* 144: 1357–1368

33 Vakeva A, Meri S (1998). Complement activation and regulator expression after anoxic injury of human endothelial cells. *APMIS* 106: 1149–1156

34 Pratt JR, Jones ME, Dong J, Zhou W, Chowdhury P, Smith RA, Sacks SH (2003) Nontransgenic hyperexpression of a complement regulator in donor kidney modulates transplant ischemia/reperfusion damage, acute rejection, and chronic nephropathy. *Am J Pathol* 163: 1457–1465

35 Arumugam TV, Shiels IA, Strachan AJ, Abbenante G, Fairlie DP, Taylor SM (2003) A small molecule C5a receptor antagonist protects kidneys from ischemia/reperfusion injury in rats. *Kidney Int* 63: 134–142

36 de Vries B, Kohl J, Leclercq WK, Wolfs TG, van Bijnen AA, Heeringa P, Buurman WA (2003) Complement factor C5a mediates renal ischemia-reperfusion injury independent from neutrophils. *J Immunol* 170: 3883–3889

37 de Vries B, Matthijsen RA, Wolfs TG, van Bijnen AA, Heeringa P, Buurman WA (2003)

Inhibition of complement factor C5 protects against renal ischemia-reperfusion injury: inhibition of late apoptosis and inflammation. *Transplantation* 75: 375–382

38 Park P, Haas M, Cunningham PN, Alexander JJ, Bao L, Guthridge JM, Kraus DM, Holers VM, Quigg RJ (2001) Inhibiting the complement system does not reduce injury in renal ischemia reperfusion. *J Am Soc Nephrol* 12: 1383–1390

39 Park P, Haas M, Cunningham PN, Bao L, Alexander JJ, Quigg RJ (2002) Injury in renal ischemia-reperfusion is independent from immunoglobulins and T lymphocytes. *Am J Physiol Renal Physiol* 282: F352–F357

40 Cooper DK (2003) Clinical xenotransplantion--how close are we? Lancet 362: 557–559

41 Lawson JH, Platt JL (1996) Molecular barriers to xenotransplantation. *Transplantation* 62: 303–310

42 Auchincloss H, Jr., Sachs DH (1998) Xenogeneic transplantation. *Annu Rev Immunol* 16: 433–470

43 Kissmeyer-Nielsen F, Olsen S, Petersen VP, Fjeldborg O (1966) Hyperacute rejection of kidney allografts, associated with pre-existing humoral antibodies against donor cells. *Lancet* 2: 662–665

44 Monteiro F, Buelow R, Mineiro C, Rodrigues H, Kalil J (1997) Identification of patients at high risk of graft loss by pre- and posttransplant monitoring of anti-HLA class I IgG antibodies by enzyme-linked immunosorbent assay. *Transplantation* 63: 542–546

45 Platt JL, Vercellotti GM, Dalmasso AP, Matas AJ, Bolman RM, Najarian JS, Bach FH (1990) Transplantation of discordant xenografts: a review of progress. *Immunol Today* 11: 450–456

46 Perper RJ, Najarian JS (1966) Experimental renal heterotransplantation. I. In widely divergent species. *Transplantation* 4: 377–388

47 Leventhal JR, Dalmasso AP, Cromwell JW, Platt JL, Manivel CJ, Bolman RM, III, Matas AJ (1993) Prolongation of cardiac xenograft survival by depletion of complement. *Transplantation* 55: 857–865

48 Miyagawa S, Shirakura R, Matsumiya G, Fukushima N, Nakata S, Matsuda H, Matsumoto M, Kitamura H, Seya T (1993) Prolonging discordant xenograft survival with anticomplement reagents K76COOH and FUT175. *Transplantation* 55: 709–713

49 Baldwin WM, III, Pruitt SK, Brauer RB, Daha MR, Sanfilippo F (1995) Complement in organ transplantation. Contributions to inflammation, injury, and rejection. *Transplantation* 59: 797–808

50 Platt JL, Fischel RJ, Matas AJ, Reif SA, Bolman RM, Bach FH (1991) Immunopathology of hyperacute xenograft rejection in a swine-to-primate model. *Transplantation* 52: 214–220

51 Leventhal JR, Matas AJ, Sun LH, Reif S, Bolman RM, III, Dalmasso AP, Platt JL (1993) The immunopathology of cardiac xenograft rejection in the guinea pig-to-rat model. *Transplantation* 56: 1–8

52 Johnston PS, Wang MW, Lim SM, Wright LJ, White DJ (1992) Discordant xenograft rejection in an antibody-free model. *Transplantation* 54: 573–576

53 Zhow XJ, Niesen N, Pawlowski I, Biesecker G, Andres G, Brentjens J, Milgram F (1990)

Prolongation of survival of discordant kidney xenografts by C6 deficiency. *Transplantation* 50: 896–898
54 Brauer RB, Baldwin WM, III, Ibrahim S, Sanfilippo F (1995) The contribution of terminal complement components to acute and hyperacute allograft rejection in the rat. *Transplantation* 59: 288–293
55 Pruitt SK, Baldwin WM, III, Sanfilippo F (1996) The role of C3a and C5a in hyperacute rejection of guinea pig-to-rat cardiac xenografts. *Transplant Proc* 28: 596
56 Rosengard AM, Cary NR, Langford GA, Tucker AW, Wallwork J, White DJ (1995) Tissue expression of human complement inhibitor, decay-accelerating factor, in transgenic pigs. A potential approach for preventing xenograft rejection. *Transplantation* 59: 1325–1333
57 Diamond LE, Quinn CM, Martin MJ, Lawson J, Platt JL, Logan JS (2001) A human CD46 transgenic pig model system for the study of discordant xenotransplantation. *Transplantation* 71: 132–142
58 Diamond LE, McCurry KR, Martin MJ, McClellan SB, Oldham ER, Platt JL, Logan JS (1996) Characterization of transgenic pigs expressing functionally active human CD59 on cardiac endothelium. *Transplantation* 61: 1241–1249
59 Zhou CY, McInnes E, Parsons N, Langford G, Lancaster R, Richards A, Pino-Chavez G, Dos Santos CG, Copeman L, Carrington C, Thompson S (2002) Production and characterization of a pig line transgenic for human membrane cofactor protein. *Xenotransplantation* 9: 183–190
60 Byrne G, McCurry K, Martin M, Platt J, Logan J (1996) Development and analysis of transgenic pigs expressing the human complement regulatory protein CD59 and DAF. *Transplant Proc* 28: 759
61 Gonzalez MM, Garcia BJ, Alonso HA, Centeno CA, Lopez PE, Vazquez ME, Mosquera RJ, Requejo I, I, Manez MR (2004) [Renal xenotransplantation from hDAF pig to baboon. Experience and review]. *Actas Urol Esp* 28: 161–174
62 Menoret S, Plat M, Blancho G, Martinat-Botte F, Bernard P, Karam G, Tesson L, Renaudin K, Guillouet P, Weill B et al (2004) Characterization of human CD55 and CD59 transgenic pigs and kidney xenotransplantation in the pig-to-baboon combination. *Transplantation* 77: 1468–1471
63 Loveland BE, Milland J, Kyriakou P, Thorley BR, Christiansen D, Lanteri MB, Regensburg M, Duffield M, French AJ, Williams L et al (2004) Characterization of a CD46 transgenic pig and protection of transgenic kidneys against hyperacute rejection in non-immunosuppressed baboons. *Xenotransplantation* 11: 171–183
64 Mathew M, Bolton WK (1988) Linear C3 deposits on the tubular basement membrane in renal allograft biopsies. *Am J Kidney Dis* 12: 121–125
65 Feucht HE, Schneeberger H, Hillebrand G, Burkhardt K, Weiss M, Riethmuller G, Land W, Albert E (1993) Capillary deposition of C4d complement fragment and early renal graft loss. *Kidney Int* 43: 1333–1338
66 Caren LD, Rosenberg LT (1965) Complement in skin grafting in mice. *Immunology* 9: 359–364

67 Marsh JE, Farmer CK, Jurcevic S, Wang Y, Carroll MC, Sacks SH (2001) The allogeneic T and B cell response is strongly dependent on complement components C3 and C4. *Transplantation* 72: 1310–1318

68 Nakashima S, Qian Z, Rahimi S, Wasowska BA, Baldwin WM, III (2002) Membrane attack complex contributes to destruction of vascular integrity in acute lung allograft rejection. *J Immunol* 169: 4620–4627

69 Pratt JR, Hibbs MJ, Laver AJ, Smith RA, Sacks SH (1996) Effects of complement inhibition with soluble complement receptor-1 on vascular injury and inflammation during renal allograft rejection in the rat. *Am J Pathol* 149: 2055–2066

70 Pratt JR, Harmer AW, Levin J, Sacks SH (1997) Influence of complement on the allospecific antibody response to a primary vascularized organ graft. *Eur J Immunol* 27: 2848–2853

71 Passwell J, Schreiner GF, Nonaka M, Beuscher HU, Colten HR (1988) Local extrahepatic expression of complement genes C3, factor B, C2, and C4 is increased in murine lupus nephritis. *J Clin Invest* 82: 1676–1684

72 Feucht HE, Zwirner J, Bevec D, Lang M, Felber E, Riethmuller G, Weiss EH (1989) Biosynthesis of complement C4 messenger RNA in normal human kidney. *Nephron* 53: 338–342

73 Zhou W, Marsh JE, Sacks SH (2001) Intrarenal synthesis of complement. *Kidney Int* 59:1227–1235

74 Welch TR, Beischel LS, Witte DP (1993) Differential expression of complement C3 and C4 in the human kidney. *J Clin Invest* 92: 1451–1458

75 Sacks SH, Zhou W, Andrews PA, Hartley B (1993) Endogenous complement C3 synthesis in immune complex nephritis. *Lancet* 342: 1273–1274

76 Andrews PA, Pani A, Zhou W, Sacks SH (1994) Local transcription of complement C3 in human allograft rejection. Evidence for a pathogenic role and correlation to histology and outcome. *Transplantation* 58: 637–640

77 Tang S, Zhou W, Sheerin NS, Vaughan RW, Sacks SH (1999) Contribution of renal secreted complement C3 to the circulating pool in humans. *J Immunol* 162: 4336–4341

78 Andrews PA, Finn JE, Lloyd CM, Zhou W, Mathieson PW, Sacks SH (1995) Expression and tissue localization of donor-specific complement C3 synthesized in human renal allografts. *Eur J Immunol* 25: 1087–1093

79 Wessels MR, Butko P, Ma M, Warren HB, Lage AL, Carroll MC (1995) Studies of group B streptococcal infection in mice deficient in complement component C3 or C4 demonstrate an essential role for complement in both innate and acquired immunity. *Proc Natl Acad Sci USA* 92: 11490–11494

80 Zhang Z, Schlachta C, Duff J, Stiller C, Grant D, Zhong R (1995) Improved techniques for kidney transplantation in mice. *Microsurgery* 16: 103–109

81 Pratt JR, Basheer SA, Sacks SH (2002) Local synthesis of complement component C3 regulates acute renal transplant rejection. *Nat Med* 8: 582–187

82 Takada M, Nadeau KC, Shaw GD, Marquette KA, Tilney NL (1997) The cytokine-

adhesion molecule cascade in ischemia/reperfusion injury of the rat kidney. Inhibition by a soluble P-selectin ligand. *J Clin Invest* 99: 2682–2690

83 Schraufstatter IU, Trieu K, Sikora L, Sriramarao P, DiScipio R (2002) Complement c3a and c5a induce different signal transduction cascades in endothelial cells. *J Immunol* 169: 2102–2110

84 Nataf S, Davoust N, Ames RS, Barnum SR (1999) Human T cells express the C5a receptor and are chemoattracted to C5a. *J Immunol* 162: 4018–4023

85 Arvieux J, Yssel H, Colomb MG (1988) Antigen-bound C3b and C4b enhance antigen-presenting cell function in activation of human T-cell clones. *Immunology* 65: 229–235

86 Erdei A, Kohler V, Schafer H, Burger R (1992) Macrophage-bound C3 fragments as adhesion molecules modulate presentation of exogenous antigens. *Immunobiology* 185: 314–326

87 Kerekes K, Prechl J, Bajtay Z, Jozsi M, Erdei A (1998) A further link between innate and adaptive immunity: C3 deposition on antigen-presenting cells enhances the proliferation of antigen-specific T cells. *Int Immunol* 10: 1923–1930

88 Li K, Patel H, Farrar CA, Hargreaves RE, Sacks SH, Zhou W (2004) Complement activation regulates the capacity of proximal tubular epithelial cell to stimulate alloreactive T cell response. *J Am Soc Nephrol* 15: 2414–2422

89 Dempsey PW, Allison ME, Akkaraju S, Goodnow CC, Fearon DT (1996) C3d of complement as a molecular adjuvant: bridging innate and acquired immunity. *Science* 271: 348–350

90 Carroll MC (2004) The complement system in regulation of adaptive immunity. *Nat Immunol* 5: 981–986

91 Racusen LC, Halloran PF, Solez K (2004) Banff 2003 meeting report: new diagnostic insights and standards. *Am J Transplant* 4: 1562–1566

92 Lederer SR, Kluth-Pepper B, Schneeberger H, Albert E, Land W, Feucht HE (2001) Impact of humoral alloreactivity early after transplantation on the long-term survival of renal allografts. *Kidney Int* 59: 334–341

93 Mauiyyedi S, Pelle PD, Saidman S, Collins AB, Pascual M, Tolkoff-Rubin NE, Williams WW, Cosimi AA, Schneeberger EE, Colvin RB (2001) Chronic humoral rejection: identification of antibody-mediated chronic renal allograft rejection by C4d deposits in peritubular capillaries. *J Am Soc Nephrol* 12: 574–582

94 Nickeleit V, Zeiler M, Gudat F, Thiel G, Mihatsch MJ (2002) Detection of the complement degradation product C4d in renal allografts: diagnostic and therapeutic implications. *J Am Soc Nephrol* 13: 242–251

95 Bohmig GA, Exner M, Habicht A, Schillinger M, Lang U, Kletzmayr J, Saemann MD, Horl WH, Watschinger B, Regele H (2002) Capillary C4d deposition in kidney allografts: a specific marker of alloantibody-dependent graft injury. *J Am Soc Nephrol* 13: 1091–1099

96 Herzenberg AM, Gill JS, Djurdjev O, Magil AB (2002) C4d deposition in acute rejection: an independent long-term prognostic factor. *J Am Soc Nephrol* 13: 234–241

97 Mauiyyedi S, Crespo M, Collins AB, Schneeberger EE, Pascual MA, Saidman SL,

Tolkoff-Rubin NE, Williams WW, Delmonico FL, Cosimi AB, Colvin RB (2002) Acute humoral rejection in kidney transplantation: II. Morphology, immunopathology, and pathologic classification. *J Am Soc Nephrol* 13: 779–787

98 Regele H, Bohmig GA, Habicht A, Gollowitzer D, Schillinger M, Rockenschaub S, Watschinger B, Kerjaschki D, Exner M (2002) Capillary deposition of complement split product C4d in renal allografts is associated with basement membrane injury in peritubular and glomerular capillaries: a contribution of humoral immunity to chronic allograft rejection. *J Am Soc Nephrol* 13: 2371–2380

99 Feucht HE (2003) Complement C4d in graft capillaries -- the missing link in the recognition of humoral alloreactivity. *Am J Transplant* 3: 646–652

100 Andrews PA, Finn JE, Mathieson PW, Sacks SH (1995) Molecular analysis of C3 allotypes related to transplant outcome in human renal allografts. *Transplantation* 60: 1342–1346

101 Takahashi K, Saito K, Takahara S, Okuyama A, Tanabe K, Toma H, Uchida K, Hasegawa A, Yoshimura N, Kamiryo Y (2004) Excellent long-term outcome of ABO-incompatible living donor kidney transplantation in Japan. *Am J Transplant* 4: 1089–1096

102 Hong Y, Zhou W, Li K, Sacks SH (2002) Triptolide is a potent suppressant of C3, CD40 and B7h expression in activated human proximal tubular epithelial cells. *Kidney Int* 62: 1291–1300

103 Zach TL, Hill LD, Herrman VA, Leuschen MP, Hostetter MK (1992) Effect of glucocorticoids on C3 gene expression by the A549 human pulmonary epithelial cell line. *J Immunol* 148: 3964–3969

104 Strunk RC, Tashjian AH, Jr., Colten HR (1975) Complement biosynthesis *in vitro* by rat hepatoma cell strains. *J Immunol* 114: 331–335

105 Lemercier C, Julen N, Coulpier M, Dauchel H, Ozanne D, Fontaine M, Ripoche J (1992) Differential modulation by glucocorticoids of alternative complement protein secretion in cells of the monocyte/macrophage lineage. *Eur J Immunol* 22: 909–915

C1q and the glomerulonephritides: therapeutic approaches for the treatment of complement-mediated kidney diseases

Stefan P. Berger, Tom W.L. Groeneveld, Anja Roos and Mohamed R. Daha

Department of Nephrology, C3-P25, Leiden University Medical Center, P.O. Box 9600, 2300 RC Leiden, The Netherlands

Introduction

Numerous forms of glomerulonephritis are associated with the deposition of immunoglobulins in the glomerulus. Immunoglobulin deposition can lead to the interaction of C1q with its ligands, e.g., IgG- or IgM-containing immune complexes. C1q is a large multimeric molecule consisting of six trimeric subunits and is associated with the serine proteases C1r and C1s in a large pro-enzymatic complex. The C1q molecule is characterized by a head and a tail subunit. The head portion is a globular domain that recognizes ligands, such as antigen-antibody complexes. The collagenous tail binds to C1q receptors and interacts with C1r and C1s. Upon activation by ligand binding, these enzymes cleave C4 and C2, leading to the formation of the classical pathway C3-convertase that may result in the release of the chemoattractive anaphylatoxins C3a and C5a, and the activation of the terminal pathway of complement. The pathological effect of antibody-mediated damage depends on the site of complement activation. In membranous nephropathy immunoglobulin deposition occurs subepithelially. The resulting co-deposition of C3 does not result in the influx of inflammatory cells, presumably because this site is not accessible for cells residing in the blood compartment. The complement products produced in this disease are most probably released in the pre-urine. If on the other hand immune complexes accumulate in a subendothelial location, products of complement activation like C5a will be released into the bloodstream and lead to chemotaxis and activation of inflammatory cells. Examples of this kind of damage include post-infectious nephritis, anti-glomerular basement membrane (GBM) nephritis and certain forms of lupus nephritis. The third site of immune complex deposition is the mesangium. It is accessible without crossing the GBM. Immune complex deposition at this site can lead to mesangial proliferation and occasionally to extracapillary proliferation.

Complement and Kidney Disease, edited by Peter F. Zipfel

C1q deposition patterns in human glomerulonephritis

Interestingly, C1q deposition is only found in a limited number of glomerulonephritides associated with immunoglobulin deposition. The "full house" pattern of immunofluorescence including IgG, IgM, IgA, C3 and C1q is the hallmark of glomerulonephritis in patients with systemic lupus erythematosus (SLE). The role of C1q and specifically that of autoantibodies against C1q in this disease is discussed later in this review. In idiopathic membranous nephritis, granular staining for IgG and C3 along the GBM is seen, whereas C1q deposition is usually not found. If C1q is detected in the presence of a morphological and immunofluorescence pattern otherwise compatible with membranous nephritis, this will lead to the suggestion of WHO class V lupus nephritis or C1q nephropathy [1, 2]. Anti-GBM disease, an inflammatory form of glomerulonephritis caused by antibodies directed against the α-3 chain of collagen type VI, is characterized by a linear deposition of IgG and C3 along the basement membrane. C1q deposition is an unusual finding in anti-GBM glomerulonephritis. In membranoproliferative glomerulonephritis type I, complement deposition is dominated by C3, though C1q deposition is regularly found. An overview of site and character of complement deposition in various forms of glomerulonephritis is summarized in Table 1.

The absence of detectable C1q in the presence of glomerular staining for IgG and C3 does not exclude that the classical pathway is responsible for the complement deposition in these kidneys. Since C1q does not undergo covalent binding to its ligands, in contrast to C4b and C3b, it may have a short half-life and can be difficult to detect. In line with this explanation, the frequent detection of C1q in lupus nephritis may be related to a local stabilizing effect of anti-C1q antibodies [3].

Complement-mediated damage in animal models of glomerulonephritis

Numerous animal models have been studied to explore the role of complement in renal damage. Early studies of antibody-induced glomerular injury in rats linked complement-associated damage to the influx of neutrophils induced by C5a [4]. Complement depletion by administration of human IgG prior to the treatment with nephrotoxic serum markedly diminished neutrophil influx and renal damage in this model. Later it was shown that depletion of the complement system is beneficial in the non-inflammatory passive Heymann nephritis model, which morphologically resembles membranous nephropathy in humans [5]. These findings suggested neutrophil-independent mechanisms of complement-mediated damage. Studies utilizing C6-depleted or C6-deficient rats more specifically demonstrated the importance of the terminal pathway of complement in various models of renal disease. In the anti-thymocyte serum model of mesangioproliferative glomerulonephritis, the absence of

Table 1 - Distribution of complement and immunoglobulins in glomerulonephritides

Disease	Glomerular deposition
Focal segmental glomerulosclerosis (FSGS)	Focal IgM and C3
Membranoproliferative glomerulonephritis I (MPGN I)	Granular C3, IgG, C1q along capillary wall and in mesangium
Membranoproliferative glomerulonephritis II (MPGN II)	C3 along capillary wall and in mesangium
IgA nephropathy	IgA, IgG, IgM, C3 and MBL in mesangium
Membranous glomerulonephropathy	Granular IgG and C3 along GBM
Post-streptococcal glomerulonephritis	Granular IgG and C3 along capillary wall and in mesangium
Goodpasture syndrome	Linear IgG and C3 and occasionally C1q in GBM pattern
Lupus nephritis	'Full house' pattern of IgG, IgM, IgA, C1q, C4 and C3 along capillary wall and in mesangium

C6 was associated with a marked reduction in various parameters of renal damage [6]. The effect was comparable to the administration of cobra venom factor. C6 depletion using an anti-C6 antibody abolished proteinuria in the passive Heymann nephritis model [7]. These data established the relative importance of C5b-9 in complement-mediated renal damage. C5b-9 has also been linked to interstitial fibrosis and inflammation in a non-immune-mediated model of proteinuric renal disease induced with the aminonucleoside puromycin [8].

C5b-9 may exert its effect in the Heymann nephritis model by increasing the formation of oxygen radicals by podocytes [9, 10]. Similarly C5b-9 has been shown to activate endothelial and mesangial cells [11, 12]. These reactive oxygen species may then cause damage to matrix proteins and lipid peroxidation. Activation of complement in proteinuric urine has been proposed as an important mediator of chronic progressive renal damage irrespective of the underlying glomerular disease [13].

Taken together, there is clear evidence for a pro-inflammatory and pathogenic role of complement activation products in renal disease. However, the pathway that is responsible for complement activation is in most cases not established. Recent data suggested a role for the lectin pathway in complement activation in IgA nephropathy [14, 15] and in lupus nephritis [16]. However, direct identification of mechanisms of complement activation has been hampered by the lack of appropriate animal models for complement-dependent renal disease.

C1q in glomerulonephritis

Few tools are available to study the relative importance of C1q in complement-mediated renal damage. As described earlier, C1q deposition is regularly found in lupus nephritis, but not in most other forms of glomerulonephritis. In humans there is a strong association between the homozygous deficiency of the early components of complement activation (C1q, C4 and C2) and the development of SLE [17]. It should be noted that the effects of C1q and C4 deficiency are much stronger than the effect of C2 deficiency. Gene-targeted C1q-deficient mice have been recently used to study the role of the classical pathway in SLE.

In mice the effect of C1q deficiency strongly depends on the genetic background. Both C1q- and C4-deficient mice with a 129_C57BL/6 genetic background develop glomerulonephritis associated with autoantibodies and accumulation of apoptotic cells [18, 19]. No evidence of glomerulonephritis was found in C1q-deficient C57BL/6 mice. Similarly, C1q deficiency does not lead to a significant change in the severity of glomerulonephritis in the MLR/*lpr* mouse [20]. Both the high incidence of SLE with severe glomerulonephritis in C1q-deficient humans, and the aggravation of lupus nephritis in certain C1q-deficient mouse strains, show that activation of the classical pathway is not an essential component in the development of renal damage in lupus nephritis. In fact, C1q seems to play a protective role in the setting of SLE. This may be explained by the role of C1q in the clearance of immune complexes [21] and apoptotic cells [22–24]. Absence of C1q may lead to defective clearance of apoptotic cells, leading to increased exposure of the immune system to autoantigens. This concept is underscored by the finding that high concentrations of the autoantigens identified in SLE are found in apoptotic blebs [25].

This beneficial role of C1q may not only apply to SLE. Robson et al. [26] studied the effect of C1q deficiency in the accelerated nephrotoxic serum model of glomerulonephritis in mice, using injection of heterologous rabbit anti-GBM antibodies in mice that were immunized with rabbit IgG. C1q-deficient mice developed more severe glomerular thrombosis compared to wild-type mice. This exacerbation of disease was associated with increased IgG deposits, enhanced influx of neutrophils and increased numbers of apoptotic cells. Defective processing of immune complexes seems to result in increased influx of neutrophils, with Fc-receptors being more important mediators of damage than complement in this model. Using a different model, in which primary injury was studied following injection of rabbit anti-mouse GBM antibodies in mice, Sheerin et al. [27] showed that the development of prompt renal injury in this model was C3 dependent. Although the classical pathway is very likely to be responsible for complement activation in this model, the role of C1q in such a model has not yet been studied.

Anti-C1q antibodies

Autoantibodies reacting with the collagenous portion of the C1q molecule [28] have been described in patients with various autoimmune diseases. In SLE, anti-C1q antibodies are detectable in 30–40% of the patients, whereas this incidence is almost 100% in the hypocomplementemic urticarial vasculitis syndrome (HUVS). In patients with SLE, the presence of anti-C1q correlates with active renal disease. Anti-C1q antibodies have been reported to correlate with lupus nephritis with a sensitivity of 87% and a specificity of 92% [29], and a rise in the anti-C1q antibody titer has been suggested to predict renal flares [29, 30]. A number of studies have shown that patients with a negative anti-C1q titer generally do not develop lupus nephritis [31]. Obviously, these data suggest an important role of anti-C1q antibodies in the development of lupus nephritis. Several animal models have been developed to study whether C1q antibodies actually are pathogenic.

Treatment of naïve mice with rabbit polyclonal antibodies directed against mouse C1q resulted in glomerular deposition of C1q and anti-C1q antibodies. Although this treatment resulted in stable deposition of C1q and anti-C1q antibodies, these mice did not develop overt glomerulonephritis [32]. The injection of anti-C1q antibodies did not result in the glomerular deposition of C1q and anti-C1q in IgG-deficient Rag2$^{-/-}$ mice [33], suggesting that minor amounts of IgG associated with the GBM are necessary for the deposition of C1q and anti-C1q in the glomerulus.

Antibodies directed against C1q have also been described in murine models of SLE. To study whether the disease associations of anti-C1q antibodies in mice are comparable with human SLE, the course of the anti-C1q titer in relation to renal and non-renal manifestations of SLE in MLR-*lpr* mouse has been described [34]. With increasing age of the mice, a rise in the anti-C1q titer and a decrease of the C1q concentration was observed. Worsening glomerulonephritis with signs of complement activation, cell influx and loss of renal function paralleled the increase in anti-C1q.

The generation of homologous mouse anti-mouse C1q antibodies provided a tool to acquire experimental evidence for a pathogenic role of C1q antibodies [3]. These antibodies were generated by immunizing C1q-deficient mice with purified mouse C1q. An antibody recognizing the tail region of C1q was used in the *in vivo* experiments. Administration of this antibody to healthy wild-type C57BL/6 mice resulted in depletion of circulating C1q and in glomerular deposition of these antibodies together with C1q. This was not accompanied by evidence of renal damage. However, if mice were pre-treated with a subnephritogenic dose of rabbit anti-GBM antibodies, administration of anti-C1q antibodies resulted in increased deposition of immunoglobulins and complement as well as strong renal damage.

Figure 1 shows increased staining for IgG and C1q in this model. These findings imply that glomerular deposition of C1q-containing immune complexes provide the

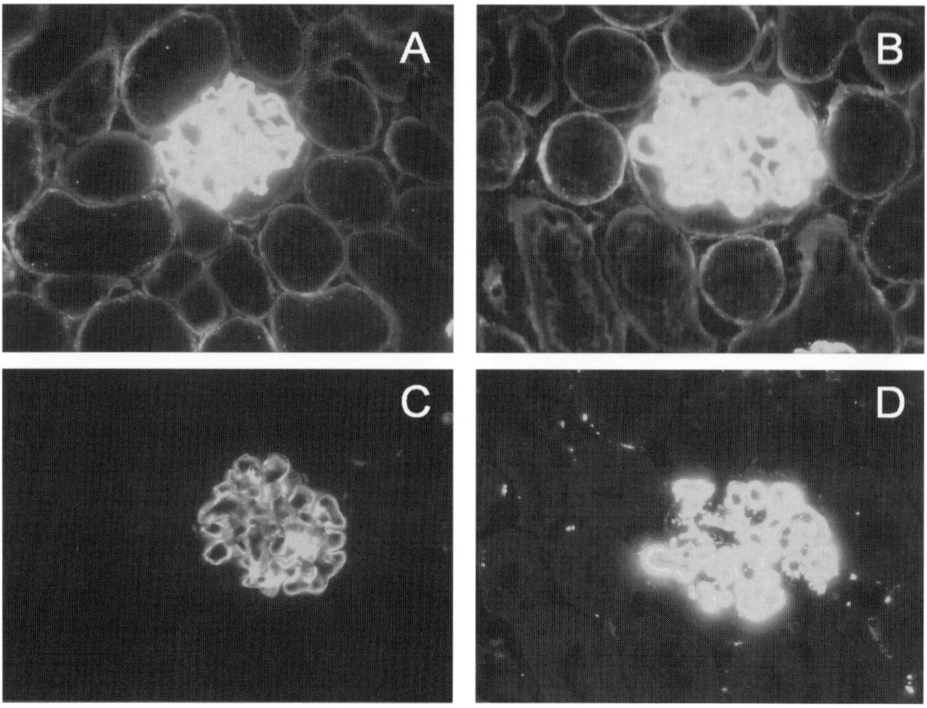

Figure 1
Anti-C1q antibodies enhance glomerular deposition of C1q
Mice (C57BL/6) received a single injection of rabbit anti-GBM antibodies alone (A, C) or followed by an mouse anti-mouse C1q antibody (B, D). After 24 h, kidney sections were stained for rabbit IgG (A, B) and C1q (C, D). Magnification × 400.

indispensable condition for anti-C1q antibodies to cause renal damage. The proposed sequence of events is illustrated in Figure 2. This model was also applied to mice genetically deficient for C4, C3 or all three Fc-γ receptors. These experiments showed that renal damage was dependent on both complement activation and Fc-γ receptor-mediated influx of inflammatory cells.

Therapeutic options

As described above, the contribution of C1q in the pathogenesis of renal disease is actually not well established. C1q deposition is detected in a limited number of renal afflictions with lupus nephritis being the most prominent one. However, from the present data it is clear that C1q plays a dual role in SLE. On one hand, C1q is

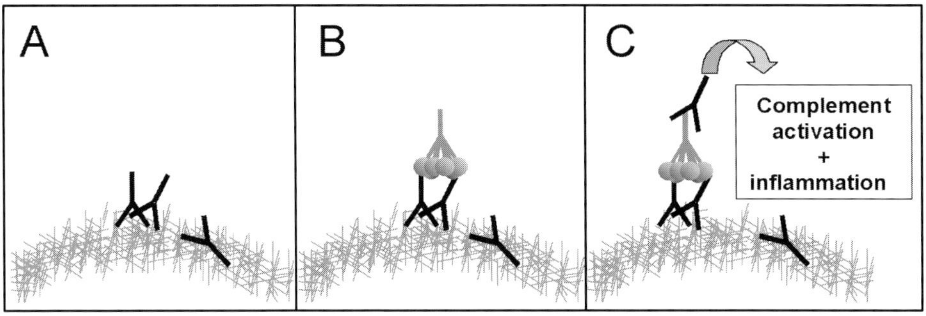

Figure 2
Proposed cascade of events in experimental lupus nephritis
(A) Immune complexes deposit or form locally in the GBM. (B) C1q molecules from the circulation bind to the immune complexes deposited in the GBM. (C) When sufficient anti-C1q autoantibodies bind the C1q attached to the immune complexes, inhibition by complement regulators will be overruled, which leads to complement activation, inflammation and subsequently renal damage.

involved in the clearance of immune complexes and apoptotic cells, and C1q deficiency is involved in a loss of self tolerance, resulting in systemic autoimmunity and lupus-like renal disease both in mice and in humans. On the other hand, the classical pathway is likely to contribute to the local antibody-mediated complement activation in lupus nephritis. As described above, recent data in a mouse model support a role for the classical pathway of complement in conjunction with anti-C1q antibodies for the development of renal injury. Nevertheless, in view of this dual role of complement and early complement proteins in SLE, there is at present no support for a complement-inhibitory approach in lupus nephritis.

Inhibition of the earliest step of classical pathway activation may offer several attractive advantages that would not be achieved by an intervention further downstream in the complement pathway. First of all, such an approach would allow innate defense via the alternative pathway and the lectin pathway to protect the individual against microbial attack. Furthermore, an early intervention would block the pro-inflammatory effects of early complement factors, such as C1q and C4a. For example, the interaction of endothelial cells with C1q leads to the production of IL-8 and IL-6 and expression of adhesion molecules [35, 36]. However, for such an approach, more research is required for the identification of the role of C1q in complement-dependent renal injury. In this respect, the generation of *C1q* gene-targeted mice and the recent availability of C1q-inhibitory monoclonal antibodies [3] allow more in depth studies in relevant experimental models.

Several options are conceivable for therapeutic inhibition of C1q (reviewed in [37]). Anti-C1q therapeutics could be directed at either inhibiting the interaction of the C1q globular head with target molecules or at impairing complement activation and interaction with complement receptors by the binding to the collagenous tail. C1q-inhibitory monoclonal antibodies and C1q-binding peptides have been described which are able to interfere in the interaction between C1q and immunoglobulin [38, 39]. Furthermore, natural C1q-binding proteins have been described that inhibit C1q function [40]. Also the natural regulatory protein C1 inhibitor could be useful in this respect, although this protein would also inhibit the lectin pathway of complement. In previous studies from our group, it has been shown that preparations of intravenous immunoglobulin, especially those that consist mainly of IgM, were effective in the inhibition of complement-mediated renal injury in an acute model of glomerulonephritis mediated by the classical complement pathway [41].

Next to the necessity of proof of principle of C1q inhibition in relevant animal models for renal disease, it should be considered whether chronic antibody-mediated renal diseases could be treated on the long-term with complement inhibition, both from a logistic and a medical point of view. In view of the clear pathological consequences of complement deficiencies, chronic complement inhibition may also confer risks for infections and autoimmunity, although this has not yet been studied. It is conceivable that such a therapy would be limited to acute and early phases of disease.

References

1 Jennette JC, Hipp CG (1985) C1q nephropathy: a distinct pathologic entity usually causing nephrotic syndrome. *Am J Kidney Dis* 6: 103–110

2 Sharman A, Furness P, Feehally J (2004) Distinguishing C1q nephropathy from lupus nephritis. *Nephrol Dial Transplant* 19: 1420–1426

3 Trouw LA, Groeneveld TW, Seelen MA, Duijs JM, Bajema IM, Prins FA, Kishore U, Salant DJ, Verbeek JS, van Kooten C et al (2004) Anti-C1q autoantibodies deposit in glomeruli but are only pathogenic in combination with glomerular C1q-containing immune complexes. *J Clin Invest* 114: 679–688

4 Cochrane CG, Unanue ER, Dixon FJ (1965) A role of polymorphonuclear leukocytes and complement in nephrotoxic nephritis. *J Exp Med* 122: 99–116

5 Salant DJ, Belok S, Madaio MP, Couser WG (1980) A new role for complement in experimental membranous nephropathy in rats. *J Clin Invest* 66: 1339–1350

6 Brandt J, Pippin J, Schulze M, Hansch GM, Alpers CE, Johnson RJ, Gordon K, Couser WG (1996) Role of the complement membrane attack complex (C5b-9) in mediating experimental mesangioproliferative glomerulonephritis. *Kidney Int* 49: 335–343

7 Baker PJ, Ochi RF, Schulze M, Johnson RJ, Campbell C, Couser WG (1989) Depletion of C6 prevents development of proteinuria in experimental membranous nephropathy in rats. *Am J Pathol* 135: 185–194

8 Nangaku M, Pippin J, Couser WG (1999) Complement membrane attack complex (C5b-9) mediates interstitial disease in experimental nephrotic syndrome. *J Am Soc Nephrol* 10: 2323–2331

9 Shah SV (1988) Evidence suggesting a role for hydroxyl radical in passive Heymann nephritis in rats. *Am J Physiol* 254: F337–F344

10 Neale TJ, Ojha PP, Exner M, Poczewski H, Ruger B, Witztum JL, Davis P, Kerjaschki D (1994) Proteinuria in passive Heymann nephritis is associated with lipid peroxidation and formation of adducts on type IV collagen. *J Clin Invest* 94: 1577–1584

11 Benzaquen LR, Nicholson-Weller A, Halperin JA (1994) Terminal complement proteins C5b-9 release basic fibroblast growth factor and platelet-derived growth factor from endothelial cells. *J Exp Med* 179: 985–992

12 Couser WG, Pippin JW, Shankland SJ (2001) Complement (C5b-9) induces DNA synthesis in rat mesangial cells *in vitro*. *Kidney Int* 59: 905-912

13 Hsu SI and Couser WG (2003) Chronic progression of tubulointerstitial damage in proteinuric renal disease is mediated by complement activation: a therapeutic role for complement inhibitors? *J Am Soc Nephrol* 14: S186–S191

14 Endo M, Ohi H, Ohsawa I, Fujita T, Matsushita M, Fujita T (1998) Glomerular deposition of mannose-binding lectin (MBL) indicates a novel mechanism of complement activation in IgA nephropathy. *Nephrol Dial Transplant* 13: 1984–1990

15 Roos A, Bouwman LH, Gijlswijk-Janssen DJ, Faber-Krol MC, Stahl GL, Daha MR (2001) Human IgA activates the complement system via the mannan-binding lectin pathway. *J Immunol* 167: 2861–2868

16 Lhotta K, Wurzner R, Konig P (1999) Glomerular deposition of mannose-binding lectin in human glomerulonephritis. *Nephrol Dial Transplant* 14: 881–886

17 Manderson AP, Botto M, Walport MJ (2004) The role of complement in the development of systemic lupus erythematosus. *Annu Rev Immunol* 22: 431–456

18 Botto M, Dell'Agnola C, Bygrave AE, Thompson EM, Cook HT, Petry F, Loos M, Pandolfi PP, Walport MJ (1998) Homozygous C1q deficiency causes glomerulonephritis associated with multiple apoptotic bodies. *Nat Genet* 19: 56–59

19 Paul E, Pozdnyakova OO, Mitchell E, Carroll MC (2002) Anti-DNA autoreactivity in C4-deficient mice. *Eur J Immunol* 32: 2672–2679

20 Mitchell DA, Pickering MC, Warren J, Fossati-Jimack L, Cortes-Hernandez J, Cook HT, Botto M, Walport MJ (2002) C1q deficiency and autoimmunity: the effects of genetic background on disease expression. *J Immunol* 168: 2538–2543

21 Nash JT, Taylor PR, Botto M, Norsworthy PJ, Davies KA, Walport MJ (2001) Immune complex processing in C1q-deficient mice. *Clin Exp Immunol* 123: 196–202

22 Taylor PR, Carugati A, Fadok VA, Cook HT, Andrews M, Carroll MC, Savill JS, Henson PM, Botto M, Walport MJ (2000) A hierarchical role for classical pathway complement proteins in the clearance of apoptotic cells *in vivo*. *J Exp Med* 192: 359–366

23 Roos A, Xu W, Castellano G, Nauta AJ, Garred P, Daha MR, van Kooten C (2004) Mini-review: A pivotal role for innate immunity in the clearance of apoptotic cells. E*ur J Immunol* 34: 921–929

24 Nauta AJ, Castellano G, Xu W, Woltman AM, Borrias MC, Daha MR, van Kooten C, Roos A (2004) Opsonization with C1q and mannose-binding lectin targets apoptotic cells to dendritic cells. *J Immunol* 173: 3044–3050

25 Casciola-Rosen LA, Anhalt G, Rosen A (1994) Autoantigens targeted in systemic lupus erythematosus are clustered in two populations of surface structures on apoptotic keratinocytes. *J Exp Med* 179: 1317–1330

26 Robson MG, Cook HT, Botto M, Taylor PR, Busso N, Salvi R, Pusey CD, Walport MJ, Davies KA (2001) Accelerated nephrotoxic nephritis is exacerbated in C1q-deficient mice. *J Immunol* 166: 6820–6828

27 Sheerin NS, Springall T, Abe K, Sacks SH (2001) Protection and injury: the differing roles of complement in the development of glomerular injury. *Eur J Immunol* 31: 1255–1260

28 Uwatoko S, Mannik M (1988) Low-molecular weight C1q-binding immunoglobulin G in patients with systemic lupus erythematosus consists of autoantibodies to the collagen-like region of C1q. *J Clin Invest* 82: 816–824

29 Moroni G, Trendelenburg M, Del Papa N, Quaglini S, Raschi E, Panzeri P, Testoni C, Tincani A, Banfi G, Balestrieri G et al (2001) Anti-C1q antibodies may help in diagnosing a renal flare in lupus nephritis. *Am J Kidney Dis* 37: 490–498

30 Marto N, Bertolaccini ML, Calabuig E, Hughes GR, Khamashta MA (2005) Anti-C1q antibodies in nephritis: correlation between titres and renal disease activity and positive predictive value in systemic lupus erythematosus. *Ann Rheum Dis* 64: 444–448. Epub 2004 Jul 29

31 Trendelenburg M, Marfurt J, Gerber I, Tyndall A, Schifferli JA (1999) Lack of occurrence of severe lupus nephritis among anti-C1q autoantibody-negative patients. *Arthritis Rheum* 42: 187–188

32 Trouw LA, Seelen MA, Duijs JM, Benediktsson H, van Kooten C, Daha MR (2003) Glomerular deposition of C1q and anti-C1q antibodies in mice following injection of antimouse C1q antibodies. *Clin Exp Immunol* 132: 32–39

33 Trouw LA, Duijs JM, van Kooten C, Daha MR (2003) Immune deposition of C1q and anti-C1q antibodies in the kidney is dependent on the presence of glomerular IgG. *Mol Immunol* 40: 595–602

34 Trouw LA, Seelen MA, Visseren R, Duijs JM, Benediktsson H, de Heer E, Roos A, van Kooten C, Daha MR (2004) Anti-C1q autoantibodies in murine lupus nephritis. *Clin Exp Immunol* 135: 41–48

35 van den Berg RH, Faber-Krol MC, Sim RB, Daha MR (1998) The first subcomponent of complement, C1q, triggers the production of IL-8, IL-6, and monocyte chemoattractant peptide-1 by human umbilical vein endothelial cells. *J Immunol* 161: 6924–6930

36 Lozada C, Levin RI, Huie M, Hirschhorn R, Naime D, Whitlow M, Recht PA, Golden B, Cronstein BN (1995) Identification of C1q as the heat-labile serum cofactor required for immune complexes to stimulate endothelial expression of the adhesion molecules E-selectin and intercellular and vascular cell adhesion molecules 1. *Proc Natl Acad Sci USA* 92: 8378–8382

37 Roos A, Ramwadhdoebe TH, Nauta AJ, Hack CE, Daha MR (2002) Therapeutic inhibition of the early phase of complement activation. *Immunobiology* 205: 595–609
38 Hoekzema R, Martens M, Brouwer MC, Hack CE (1988) The distortive mechanism for the activation of complement component C1 supported by studies with a monoclonal antibody against the "arms" of C1q. *Mol Immunol* 25: 485–494
39 Roos A, Nauta AJ, Broers D, Faber-Krol MC, Trouw LA, Drijfhout JW, Daha MR (2001) Specific inhibition of the classical complement pathway by C1q-binding peptides. *J Immunol* 167: 7052–7059
40 Roos A, Ramwadhdoebe TH, Nauta AJ, Hack CE, Daha MR (2002) Therapeutic inhibition of the early phase of complement activation. *Immunobiology* 205: 595–609
41 Rieben R, Roos A, Muizert Y, Tinguely C, Gerritsen AF, Daha MR (1999) Immunoglobulin M-enriched human intravenous immunoglobulin prevents complement activation *in vitro* and *in vivo* in a rat model of acute inflammation. *Blood* 93: 942–951

Complement deficient mice as model systems for kidney diseases

Joshua M. Thurman and V. Michael Holers

Division of Rheumatology, B-115, University of Colorado Health Sciences Center, 4200 East 9th Avenue, Denver, CO 80262, USA

Introduction

The development of mice deficient in one or more of the components of the complement system has facilitated our ability to examine the role of complement in various disease models. As with all studies utilizing mice, those that utilize complement-deficient mice must be regarded with some skepticism. Mice that under- or overexpress various complement components may have compensatory changes in the expression of other components [1]. Furthermore, there are a number of important differences between the murine and human immune systems [2], complicating the extrapolation from mouse models to human disease.

In spite of these caveats, complement-deficient mice are a powerful tool with which to investigate the role of complement in renal disease. As discussed here, mice deficient in the activation pathways develop significantly milder renal disease in a number of different models, strongly implicating complement activation in the pathogenesis of disease. Several nephrologists have proposed that complement inhibitors should effectively treat various forms of renal disease [3, 4], and pre-clinical studies utilizing complement deficient mice are critical to the establishment of complement inhibition as a viable therapy for renal disease.

Mouse strains with targeted deletion of complement components

Targeted disruption of numerous complement genes has been performed, and some rodent strains have been identified with spontaneous inactivating mutations (Fig. 1). Gene-targeted mice now exist with selective disruption of the activation pathways, regulators of complement activation (RCAs), and complement receptors. These mice have been effectively used to study the role of complement activation in various models, and several studies have compared responses among strains deficient in different complement proteins to compare and contrast the specific roles of complement components [5–7] in the pathogenesis of disease. For example, studies have compared disease severity in C3-deficient and C4-deficient mice to determine the

Complement and Kidney Disease, edited by Peter F. Zipfel

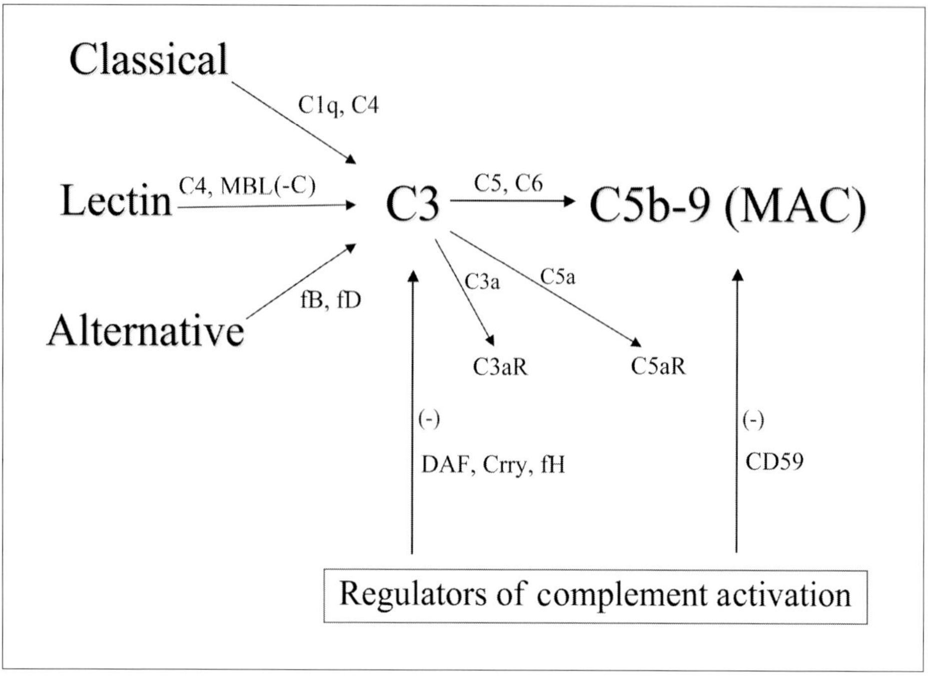

Figure 1
Complement components that have been targeted for germ-line disruption in mice. In this scheme of the complement system, components that have been selectively disrupted in existing mouse strains are indicated in red. Targeted components include factors necessary for each of the three activation pathways, the anaphylatoxin receptors, and inhibitors of complement activation.

relative contributions of the classical and alternative pathways to renal ischemia/reperfusion (I/R) injury [7]. Other strains have been used to corroborate previous results. For example, Factor D-deficient mice have been used to confirm that the results in Factor B-deficient mice are the result of a non-functioning alternative pathway [8, 9], and not some other unforeseen effect of Factor B or a linked gene that was also affected during the gene-targeting process.

Mice lacking essential components of the activation cascades are useful for studying the role of complement activation in particular disease models. For example, C3-deficient mice, which have been created by two laboratories [10, 11], effectively block activation by all three pathways. In addition, C4-deficient mice have been created that lack activation of C3 by both the classical and lectin pathways [12]. Gene-targeted mice have been generated to selectively target the classical [13], alternative [14, 15], and lectin pathways [16].

Mice deficient in downstream components of the complement system have permitted investigators to study the contribution of specific complement-activation products to the pathogenesis of disease. Specific disruption of the C3a and C5a anaphylatoxin receptors has been accomplished [17, 18], allowing specific investigation of the downstream effects of these moieties. A mutation-prone mouse strain with a deficiency of C6 has also been identified [19], rendering these mice unable to form the membrane attack complex (MAC), but permitting activation of the different pathways and generation of the C3a and C5a anaphylatoxins. Experiments with these mice can reveal the inflammatory effects of the MAC without interfering with anaphylatoxin function.

Mice that are deficient in the various RCAs have been developed. The three membrane-bound complement inhibitors expressed within the mouse kidney are decay-accelerating factor (DAF), CD59, and complement receptor 1 (CR1)-related gene y (Crry). Each of these genes has individually been targeted [20–23], and mice deficient in more than one of these inhibitors have also been studied [22]. The soluble inhibitor of alternative pathway complement activation, Factor H, has also been disrupted [24]. Autologous injury to host tissues by complement activation is accentuated in mice lacking one or more of the complement regulators, highlighting the importance of endogenous complement inhibition in regulating the inflammatory response [24–27].

Phenotypes of complement-deficient mice

Similar to humans, mice deficient in classical pathway components may be susceptible to bacterial infection [10]. Mice with C4 deficiency also have an impaired T-dependent humoral immune response [12]. Deficiency in classical pathway components may exacerbate the loss of self tolerance in models of autoimmunity, and C1q-deficient mice (which have abnormal clearance of apoptotic bodies) develop spontaneous autoimmune disease [13]. Although comparison of C3- and C4-deficient mice suggests some role for the alternative pathway in antibody-mediated clearance of some bacteria [10], factor B-deficient mice have not been demonstrated to have greater susceptibility to bacterial infection [14]. Mice deficient in the C5a receptor have impaired clearance of intrapulmonary *Pseudomonas aeruginosa* [17], and have diminished intraperitoneal and intradermal reverse passive Arthus reactions compared to wild-type controls [16].

Homozygous deficiency of Crry is lethal *in utero*; thus only mice with heterozygous deficiency of this protein survive [20]. This finding revealed the importance controlling complement activation within the placenta. Even in wild-type mice the alternative pathway appears to be activated at a low level within the placenta [28], and clearly control of complement activation is critical to the survival of the fetus. Activation of the alternative pathway also occurs within the normal renal tubuloin-

Although I/R injury of the heart, intestine, and skeletal muscle activates complement through the classical pathway, renal I/R injury appears to be unique in that complement activation occurs predominantly (if not exclusively) via the alternative pathway [7, 29]. Furthermore, there is evidence of alternative pathway activation along the tubular basement membrane of unmanipulated mice [29]. The only membrane-bound complement inhibitor expressed by the proximal tubular epithelium is Crry, and Crry is polarized to the basolateral aspect of the tubular epithelial cells (unpublished observation). Alternative pathway activation along the tubular basement membrane must therefore overwhelm the inhibitory activity of Crry at this location.

Studies in wild-type mice have revealed that the expression and localization of Crry by proximal tubular epithelial cells is rapidly altered after I/R injury. Loss of inhibition by Crry at the basolateral surface of the epithelium, therefore, renders these cells deficient of any membrane-bound complement inhibitor, permitting activation of the alternative pathway on the cell surface. Studies with immortalized proximal tubular epithelial cells have also shown that treatment with an inhibitor of Crry in the presence of mouse serum induces injury to the cells in an alternative pathway-dependent manner. Consistent with these findings, mice that are heterozygous deficient for Crry are more sensitive to mild ischemic insults (unpublished observation).

Interestingly, mice that are deficient in DAF or doubly deficient for DAF and CD59 are also more susceptible to ischemic acute renal failure (ARF) [25]. Complement activation in these mice occurs intravascularly [25], as would be expected by the localization of these proteins. In contrast, wild-type mice do not demonstrate intravascular complement deposition after I/R [7]. These findings in DAF-, DAF/CD59-, and Crry-deficient mice reveal that the different structures in the kidney are critically dependent upon the different membrane-bound RCAs after I/R. I/R renders the proximal tubular epithelium effectively deficient in expression of Crry; however, it allows alternative pathway activation to occur. Studies in complement-deficient mice have therefore helped in the discovery of this novel means of complement activation after I/R.

Studies in complement-deficient mice have also helped to reveal the mechanisms of complement-mediated injury after I/R. A study by Zhou et al. [7] demonstrated that mice deficient in C6 (and therefore unable to form the MAC) were protected from ischemic ARF. Administration of C5a receptor antagonists has also been shown to protect rodents from ischemic ARF [51, 52]. Thus, both the membrane attack complex and the anaphylatoxin C5a likely contribute to renal inflammation after I/R.

Renal transplantation

A series of elegant studies by the Sacks laboratory has demonstrated that synthesis of complement components by the kidney after allogeneic transplantation in rodents contributes to loss of the allograft [54–57] (see also the chapter by Zhou and Sacks). Kidneys from C3-deficient mice transplanted into wild-type mice showed better function than wild-type kidneys transplanted into C3-deficient recipients [56]. This finding specifically implicates C3 synthesized by the allograft in post-transplant injury to the kidney.

Complement activation may adversely affect transplant outcomes by several different mechanisms. As discussed above, complement deficiency reduces I/R injury incurred during the transplantation. The allospecific IgG response to transplanted tissue may depend on an intact classical pathway [55]. Complement activation may also influence the response of T cells to the allogeneic tissue [54]. Although many questions remain, these studies highlight the important role of complement in modulating the adaptive immune response. Given this interdependence of the complement system and the adaptive immune response, studies of immune-mediated injury in which complement-deficient mice are protected must be interpreted cautiously. Activation of complement influences the adaptive immune response, the solubilization and clearance of ICs, and the effector function of both innate and acquired immunity.

IC-mediated glomerulonephritis

The suspicion that complement activation contributes to the development of IC-mediated glomerular disease is founded upon the detection of complement-activation products in the glomeruli of affected individuals, perturbations in systemic levels of complement components, and the protection of complement-deficient animals in IC disease models [35]. ICs may deposit in the subendothelial, mesangial, and subepithelial spaces. The resulting injury depends upon the localization of the ICs within the glomerulus and the access of complement-activation fragments to the blood stream [58].

Complement-deficient mice have been tested in numerous models of antibody-mediated glomerular disease [5, 59]. The complement dependence of acute (heterologous) nephrotoxic nephritis illustrates that complement activation likely has an important role as a direct effector of antibody-mediated renal injury [6, 60]. Comparison of C4- and C3-deficient mice suggests an important role for the alternative pathway [6]. As in lupus, then, the relative protection offered by mouse strains with deletions of classical and alternative pathway components has revealed that the alternative pathway may be more important to the development of injury than traditionally thought [61].

DAF-deficient mice develop disease with subnephritogenic doses of the nephrotoxic serum [27], and they develop more severe disease than wild-type controls with nephritogenic doses of nephrotoxic serum [26]. Thus, although endogenous inhibitors of complement are overwhelmed in this and other models of IC-mediated injury, the development of disease is clearly more severe in their absence.

Models of IC-mediated glomerulonephritis in which the immune response of the mouse is important to the development of disease have shown that the classical pathway is not simply an effector mechanism of injury [5, 62, 63]. In these models, as exemplified by models of lupus nephritis, complement likely has both pro- and anti-inflammatory activity (see above). Traditional models of glomerulonephritis that emphasize the injurious consequences of complement activation on the surface of resident glomerular cells [64] must, therefore, be modified to incorporate the protean effects of complement activation, including its role in IC clearance and the adaptive immune response (Fig. 2).

In the accelerated form of (autologous) nephrotoxic nephritis, the nature of the glomerular injury and the role of complement depend upon the mouse strain used [65]. This likely reflects the tendency of a given strain to mount a Th1 or Th2 immune response [65]. An intact classical pathway may help clear apoptotic bodies and ICs in this model, and C1q-deficient mice develop more severe disease than wild-type controls [66]. H2-Bf/C2-deficient mice did not show this exacerbation of disease, suggesting that it is proteins proximal to C2 that are responsible for protection from IC-mediated injury [66], although it is possible that concurrent deficiency of the alternative pathway was protective and masked the effect of C2 deficiency. Increased IgG immune deposits as well as increased numbers of apoptotic cells were seen in the glomeruli of the C1q-deficient mice, suggesting that impaired clearance of these entities was the cause of more severe disease. Another study that examined the effects of C3 deficiency on the development of disease after injection with nephrotoxic serum demonstrated that C3-deficient mice were protected from injury in the early (heterologous) phase of injury, but were more susceptible than wild-type controls in the later (autologous) phase of injury [62]. Thus, in the heterologous phase, complement presumably functions primarily as a proinflammatory mediator of injury. As the mouse mounts an immune response to the planted antigen, however, the anti-inflammatory functions of the complement system predominate.

Progressive renal disease

Injury to the renal tubulointerstitium in progressive renal disease may occur through the filtration of complement components into the tubular lumen [67], through local synthesis of complement components [68, 69], or through activation by increased interstitial concentrations of ammonia in diseases causing reduced renal mass [70]. Because injury by these mechanisms should occur independently of IC activation of complement, they are best tested by non-IC models of progressive renal injury. Available models, such as the 5/6 nephrectomy and aminonucleoside nephrosis, are most commonly performed in rats [71, 72]. The refinement of such models in mice should permit an improved evaluation of the role of complement in progressive renal disease.

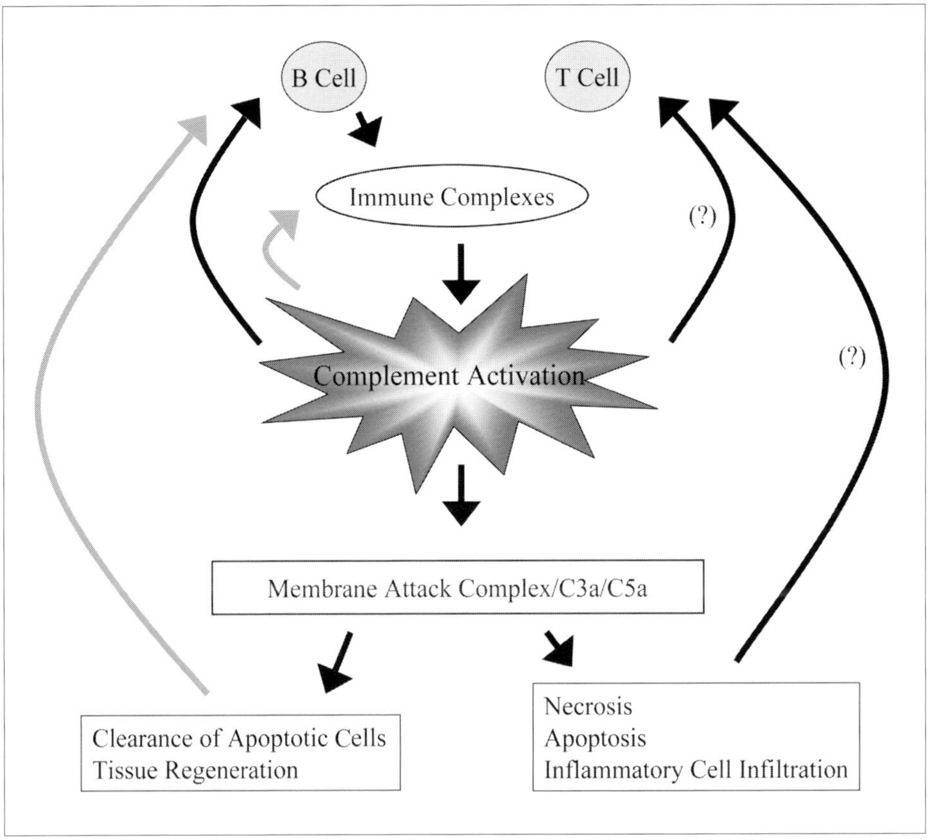

Figure 2
Role of complement in IC-mediated tissue injury. IC activation of the complement system results in the generation of pro-inflammatory complement-activation fragments. Complement activation also functions to solubilize IC and clear apoptotic cells (grey arrows indicate that these functions are anti-inflammatory). Complement-activation fragments also modulate the adaptive immune response of B cells and possibly T cells.

Conclusions

Complement-deficient mice, particularly those lacking factors necessary for the activation pathways, have demonstrated that complement activation contributes to the development of disease in several models. Complement-deficient mice have shown protection in models of IC glomerulonephritis, progressive renal disease, renal I/R injury, as well as mismatched allogeneic renal transplantation. The unexpected importance of the alternative pathway in the development of lupus nephritis underscores the importance of *in vivo* experiments utilizing multiple available knockout strains.

Mice deficient in the various RCAs have demonstrated the importance of these proteins in preventing spontaneous renal injury [24], as well as greater susceptibility to complement-mediated injury [27]. These studies confirm the necessity for constant inhibition by the host to prevent complement-mediated injury either spontaneously or as the result of bystander injury. To the degree that all complement-mediated injury to the host is a "failure" of RCAs, future studies should address the mechanisms whereby these proteins are overwhelmed.

The studies examining the role of complement in the development of renal disease in lupus-prone mice demonstrate the complexities of complement function. Models in which complement deficiency has proven protective (or not protective) may warrant detailed investigation using several of the available knockout strains to confirm a purely pro-inflammatory role for complement activation in the development of disease. Furthermore, our emerging understanding of the importance of the complement system in the clearance of apoptotic cells and in tissue regeneration [73] means that the numerous available knockouts (as well as a growing list of exogenous complement inhibitors) will be necessary to define all of the downstream consequences of complement activation. It seems likely that complement activation occurs and triggers both inflammatory and anti-inflammatory complement-dependent processes in most models in which this activation occurs.

References

1 Kang HJ, Bao L, Xu Y, Quigg RJ, Giclas PC, Holers VM (2004) Increased serum C3 levels in Crry transgenic mice partially abrogates its complement inhibitory effects. *Clin Exp Immunol* 136: 194–199

2 Mestas J, Hughes CC (2004) Of mice and not men: differences between mouse and human immunology. *J Immunol* 172: 2731–2738

3 Couser WG (2003) Complement inhibitors and glomerulonephritis: are we there yet? *J Am Soc Nephrol* 14: 815–818

4 Quigg RJ (1999) We need to inhibit complement in glomerular proteinuria. *Kidney Int* 56: 2314–2315

5 Quigg RJ, Lim A, Haas M, Alexander JJ, He C, Carroll MC (1998) Immune complex glomerulonephritis in C4- and C3-deficient mice. *Kidney Int* 53: 320–330

6 Sheerin NS, Springall T, Carroll MC, Hartley B, Sacks SH (1997) Protection against anti-glomerular basement membrane (GBM)-mediated nephritis in C3- and C4-deficient mice. *Clin Exp Immunol* 110: 403–409

7 Zhou W, Farrar CA, Abe K, Pratt JR, Marsh JE, Wang Y, Stahl GL, Sacks SH (2000)

Predominant role for C5b-9 in renal ischemia/reperfusion injury. *J Clin Invest* 105: 1363–1371
8 Elliott MK, Jarmi T, Ruiz P, Xu Y, Holers VM, Gilkeson GS (2004) Effects of complement factor D deficiency on the renal disease of MRL/lpr mice. *Kidney Int* 65: 129–138
9 Watanabe H, Garnier G, Circolo A, Wetsel RA, Ruiz P, Holers VM, Boackle SA, Colten HR, Gilkeson GS (2000) Modulation of renal disease in MRL/lpr mice genetically deficient in the alternative complement pathway factor B. *J Immunol* 164: 786–794
10 Wessels MR, Butko P, Ma M, Warren HB, Lage AL, Carroll MC (1995) Studies of group B streptococcal infection in mice deficient in complement component C3 or C4 demonstrate an essential role for complement in both innate and acquired immunity. *Proc Natl Acad Sci USA* 92: 11490–11494
11 Circolo A, Garnier G, Fukuda W, Wang X, Hidvegi T, Szalai AJ, Briles DE, Volanakis JE, Wetsel RA, Colten HR (1999) Genetic disruption of the murine complement C3 promoter region generates deficient mice with extrahepatic expression of C3 mRNA. *Immunopharmacology* 42: 135–149
12 Fischer MB, Ma M, Goerg S, Zhou X, Xia J, Finco O, Han S, Kelsoe G, Howard RG, Rothstein TL et al (1996) Regulation of the B cell response to T-dependent antigens by classical pathway complement. *J Immunol* 157: 549–556
13 Botto M (1998) C1q knock-out mice for the study of complement deficiency in autoimmune disease. *Exp Clin Immunogenet* 15: 231–234
14 Matsumoto M, Fukuda W, Circolo A, Goellner J, Strauss-Schoenberger J, Wang X, Fujita S, Hidvegi T, Chaplin DD, Colten HR (1997) Abrogation of the alternative complement pathway by targeted deletion of murine factor B. *Proc Natl Acad Sci USA* 94: 8720–8725
15 Xu Y, Ma M, Ippolito GC, Schroeder HW Jr, Carroll MC, Volanakis JE (2001) Complement activation in factor D-deficient mice. *Proc Natl Acad Sci USA* 98: 14577–14582
16 Shi L, Takahashi K, Dundee J, Shahroor-Karni S, Thiel S, Jensenius JC, Gad F, Hamblin MR, Sastry KN, Ezekowitz RA (2004) Mannose-binding lectin-deficient mice are susceptible to infection with *Staphylococcus aureus*. *J Exp Med* 199: 1379–1390
17 Hopken UE, Lu B, Gerard NP, Gerard C (1996) The C5a chemoattractant receptor mediates mucosal defence to infection. *Nature* 383: 86–89
18 Kildsgaard J, Hollmann TJ, Matthews KW, Bian K, Murad F, Wetsel RA (2000) Cutting edge: targeted disruption of the C3a receptor gene demonstrates a novel protective anti-inflammatory role for C3a in endotoxin-shock. *J Immunol* 165: 5406–5409
19 Orren A, Wallance ME, Horbart MJ, Lachmann PJ (1989) C6 polymorphism and C6 deficiency in site strains of the mutation-prone Peru-Coppock mice. *Complement Inflamm* 6: 295–296 (abstr)
20 Xu C, Mao D, Holers VM, Palanca B, Cheng AM, Molina H (2000) A critical role for murine complement regulator crry in fetomaternal tolerance. *Science* 287: 498–501
21 Lin F, Fukuoka Y, Spicer A, Ohta R, Okada N, Harris CL, Emancipator SN, Medof ME (2001) Tissue distribution of products of the mouse decay-accelerating factor (DAF)

genes. Exploitation of a Daf1 knock-out mouse and site-specific monoclonal antibodies. *Immunology* 104: 215–225

22 Miwa T, Zhou L, Hilliard B, Molina H, Song WC (2002) Crry, but not CD59 and DAF, is indispensable for murine erythrocyte protection *in vivo* from spontaneous complement attack. *Blood* 99: 3707–3716

23 Holt DS, Botto M, Bygrave AE, Hanna SM, Walport MJ, Morgan BP (2001) Targeted deletion of the CD59 gene causes spontaneous intravascular hemolysis and hemoglobinuria. *Blood* 98: 442–449

24 Pickering MC, Cook HT, Warren J, Bygrave AE, Moss J, Walport MJ, Botto M (2002) Uncontrolled C3 activation causes membranoproliferative glomerulonephritis in mice deficient in complement factor H. *Nat Genet* 31: 424–428

25 Yamada K, Miwa T, Liu J, Nangaku M, Song WC (2004) Critical Protection from Renal Ischemia Reperfusion Injury by CD55 and CD59. *J Immunol* 172: 3869–3875

26 Lin F, Emancipator SN, Salant DJ, Medof ME (2002) Decay-accelerating factor confers protection against complement-mediated podocyte injury in acute nephrotoxic nephritis. *Lab Invest* 82: 563–569

27 Sogabe H, Nangaku M, Ishibashi Y, Wada T, Fujita T, Sun X, Miwa T, Madaio MP, Song WC (2001) Increased susceptibility of decay-accelerating factor deficient mice to anti-glomerular basement membrane glomerulonephritis. *J Immunol* 167: 2791–2797

28 Girardi G, Berman J, Redecha P, Spruce L, Thurman JM, Kraus D, Hollmann TJ, Casali P, Caroll MC, Wetsel RA et al (2003) Complement C5a receptors and neutrophils mediate fetal injury in the antiphospholipid syndrome. *J Clin Invest* 112: 1644–1654

29 Thurman JM, Ljubanovic D, Edelstein CL, Gilkeson GS, Holers VM (2003) Lack of a functional alternative complement pathway ameliorates ischemic acute renal failure in mice. *J Immunol* 170: 1517–1523

30 Li B, Sallee C, Dehoff M, Foley S, Molina H, Holers VM (1993) Mouse Crry/p65. Characterization of monoclonal antibodies and the tissue distribution of a functional homologue of human MCP and DAF. *J Immunol* 151: 4295–4305

31 Qin X, Miwa T, Aktas H, Gao M, Lee C, Qian YM, Morton CC, Shahsafaei A, Song WC, Halperin JA (2001) Genomic structure, functional comparison, and tissue distribution of mouse Cd59a and Cd59b. *Mamm Genome* 12: 582–589

32 Nomura A, Nishikawa K, Yuzawa Y, Okada H, Okada N, Morgan BP, Piddlesden SJ, Nadai M, Hasegawa T, Matsuo S (1995) Tubulointerstitial injury induced in rats by a monoclonal antibody that inhibits function of a membrane inhibitor of complement. *J Clin Invest* 96: 2348–2356

33 Levy M, Halbwachs-Mecarelli L, Gubler MC, Kohout G, Bensenouci A, Niaudet P, Hauptmann G, Lesavre P (1986) H deficiency in two brothers with atypical dense intramembranous deposit disease. *Kidney Int* 30: 949–956

34 Hogasen K, Jansen JH, Mollnes TE, Hovdenes J, Harboe M (1995) Hereditary porcine membranoproliferative glomerulonephritis type II is caused by factor H deficiency. *J Clin Invest* 95: 1054–1061

35 Volanakis JE, Frank MM (eds) (1998) *The Human Complement System in Health and*

Disease. New York: Marcel Dekker, Inc.

36 Manderson AP, Botto M, Walport MJ (2004) The role of complement in the development of systemic lupus erythematosus. *Annu Rev Immunol* 22: 431–456

37 Takahashi M, Tack BF, Nussenzweig V (1977) Requirements for the solubilization of immune aggregates by complement: assembly of a factor B-dependent C3-convertase on the immune complexes. *J Exp Med* 145: 86–100

38 Vivanco F, Munoz E, Vidarte L, Pastor C (1999) The covalent interaction of C3 with IgG immune complexes. *Mol Immunol* 36: 843–852

39 Pickering MC, Botto M, Taylor PR, Lachmann PJ, Walport MJ (2000) Systemic lupus erythematosus, complement deficiency, and apoptosis. *Adv Immunol* 76: 227–324

40 Dragon-Durey MA, Quartier P, Fremeaux-Bacchi V, Blouin J, de Barace C, Prieur AM, Weiss L, Fridman WH (2001) Molecular basis of a selective C1s deficiency associated with early onset multiple autoimmune diseases. *J Immunol* 166: 7612–7616

41 Rupert KL, Moulds JM, Yang Y, Arnett FC, Warren RW, Reveille JD, Myones BL, Blanchong CA, Yu CY (2002) The molecular basis of complete complement C4A and C4B deficiencies in a systemic lupus erythematosus patient with homozygous C4A and C4B mutant genes. *J Immunol* 169: 1570–1578

42 Walport MJ, Lachmann PJ (1988) Erythrocyte complement receptor type 1, immune complexes, and the rheumatic diseases. *Arthritis Rheum* 31: 153–158

43 Botto M, Dell'Agnola C, Bygrave AE, Thompson EM, Cook HT, Petry F, Loos M, Pandolfi PP, Walport MJ (1998) Homozygous C1q deficiency causes glomerulonephritis associated with multiple apoptotic bodies. *Nat Genet* 19: 56–59

44 Mitchell DA, Pickering MC, Warren J, Fossati-Jimack L, Cortes-Hernandez J, Cook HT, Botto M, Walport MJ (2002) C1q deficiency and autoimmunity: the effects of genetic background on disease expression. *J Immunol* 168: 2538–2543

45 Prodeus AP, Goerg S, Shen LM, Pozdnyakova OO, Chu L, Alicot EM, Goodnow CC, Carroll MC (1998) A critical role for complement in maintenance of self-tolerance. *Immunity* 9: 721–731

46 Einav S, Pozdnyakova OO, Ma M, Carroll MC (2002) Complement C4 is protective for lupus disease independent of C3. *J Immunol* 168: 1036–1041

47 Sekine H, Reilly CM, Molano ID, Garnier G, Circolo A, Ruiz P, Holers VM, Boackle SA, Gilkeson GS (2001) Complement component C3 is not required for full expression of immune complex glomerulonephritis in MRL/lpr mice. *J Immunol* 166: 6444–6451

48 Bao L, Haas M, Boackle SA, Kraus DM, Cunningham PN, Park P, Alexander JJ, Anderson RK, Culhane K, Holers VM et al (2002) Transgenic expression of a soluble complement inhibitor protects against renal disease and promotes survival in MRL/lpr mice. *J Immunol* 168: 3601–3607

49 Bao L, Zhou J, Holers VM, Quigg RJ (2003) Excessive matrix accumulation in the kidneys of MRL/lpr lupus mice is dependent on complement activation. *J Am Soc Nephrol* 14: 2516–2525

50 De Vries B, Matthijsen RA, Van Bijnen AA, Wolfs TG, Buurman WA (2003) Lysophos-

phatidic Acid prevents renal ischemia-reperfusion injury by inhibition of apoptosis and complement activation. *Am J Pathol* 163: 47–56
51 De Vries B, Kohl J, Leclercq WK, Wolfs TG, Van Bijnen AA, Heeringa P, Buurman WA (2003) Complement factor C5a mediates renal ischemia-reperfusion injury independent from neutrophils. *J Immunol* 170: 3883–3889
52 Arumugam TV, Shiels IA, Strachan AJ, Abbenante G, Fairlie DP, Taylor SM (2003) A small molecule C5a receptor antagonist protects kidneys from ischemia/reperfusion injury in rats. *Kidney Int* 63: 134–142
53 Park P, Haas M, Cunningham PN, Alexander JJ, Bao L, Guthridge JM, Kraus DM, Holers VM, Quigg RJ (2001) Inhibiting the complement system does not reduce injury in renal ischemia reperfusion. *J Am Soc Nephrol* 12: 1383–1390
54 Pratt JR, Jones ME, Dong J, Zhou W, Chowdhury P, Smith RA, Sacks SH (2003) Non-transgenic hyperexpression of a complement regulator in donor kidney modulates transplant ischemia/reperfusion damage, acute rejection, and chronic nephropathy. *Am J Pathol* 163: 1457–1465
55 Marsh JE, Farmer CK, Jurcevic S, Wang Y, Carroll MC, Sacks SH (2001) The allogeneic T and B cell response is strongly dependent on complement components C3 and C4. *Transplantation* 72: 1310–1318
56 Pratt JR, Basheer SA, Sacks SH (2002) Local synthesis of complement component C3 regulates acute renal transplant rejection. *Nat Med* 8: 582–587
57 Pratt JR, Abe K, Miyazaki M, Zhou W, Sacks SH (2000) *In situ* localization of C3 synthesis in experimental acute renal allograft rejection. *Am J Pathol* 157: 825–831
58 Trouw LA, Seelen MA, Daha MR (2003) Complement and renal disease. *Mol Immunol* 40: 125–134
59 Quigg RJ (2003) Complement and the kidney. *J Immunol* 171: 3319–3324
60 Quigg RJ, He C, Lim A, Berthiaume D, Alexander JJ, Kraus D, Holers VM (1998) Transgenic mice overexpressing the complement inhibitor crry as a soluble protein are protected from antibody-induced glomerular injury. *J Exp Med* 188: 1321–1331
61 Holers VM, Thurman JM (2004) The alternative pathway of complement in disease: opportunities for therapeutic targeting. *Mol Immunol* 41: 147–152
62 Sheerin NS, Springall T, Abe K, Sacks SH (2001) Protection and injury: the differing roles of complement in the development of glomerular injury. *Eur J Immunol* 31: 1255–1260
63 Muhlfeld AS, Segerer S, Hudkins K, Farr AG, Bao L, Kraus D, Holers VM, Quigg RJ, Alpers CE (2004) Overexpression of complement inhibitor Crry does not prevent cryoglobulin-associated membranoproliferative glomerulonephritis. *Kidney Int* 65: 1214–1223
64 Couser WG (1985) Mechanisms of glomerular injury in immune-complex disease. *Kidney Int* 28: 569–583
65 Huang XR, Holdsworth SR, Tipping PG (1997) Th2 responses induce humorally mediated injury in experimental anti-glomerular basement membrane glomerulonephritis. *J Am Soc Nephrol* 8: 1101–1108

66 Robson MG, Cook HT, Botto M, Taylor PR, Busso N, Salvi R, Pusey CD, Walport MJ, Davies KA (2001) Accelerated nephrotoxic nephritis is exacerbated in C1q-deficient mice. *J Immunol* 166: 6820–6828
67 Sheerin NS, Sacks SH (2002) Leaked protein and interstitial damage in the kidney: is complement the missing link? *Clin Exp Immunol* 130: 1–3
68 Tang S, Lai KN, Chan TM, Lan HY, Ho SK, Sacks SH (2001) Transferrin but not albumin mediates stimulation of complement C3 biosynthesis in human proximal tubular epithelial cells. *Am J Kidney Dis* 37: 94–103
69 Tang S, Sheerin NS, Zhou W, Brown Z, Sacks SH (1999) Apical proteins stimulate complement synthesis by cultured human proximal tubular epithelial cells. *J Am Soc Nephrol* 10: 69–76
70 Nath KA, Hostetter MK, Hostetter TH (1985) Pathophysiology of chronic tubulo-interstitial disease in rats. Interactions of dietary acid load, ammonia, and complement component C3. *J Clin Invest* 76: 667–675
71 Rangan GK, Pippin JW, Couser WG (2004. C5b-9 regulates peritubular myofibroblast accumulation in experimental focal segmental glomerulosclerosis. *Kidney Int* 66: 1838–1848
72 Abbate M, Zoja C, Rottoli D, Corna D, Perico N, Bertani T, Remuzzi G (1999) Antiproteinuric therapy while preventing the abnormal protein traffic in proximal tubule abrogates protein- and complement-dependent interstitial inflammation in experimental renal disease. *J Am Soc Nephrol* 10: 804–813
73 Markiewski MM, Mastellos D, Tudoran R, DeAngelis RA, Strey CW, Franchini S, Wetsel RA, Erdei A, Lambris JD (2004) C3a and C3b activation products of the third component of complement (C3) are critical for normal liver recovery after toxic injury. *J Immunol* 173: 747–754

Non-Shiga toxin-associated hemolytic uremic syndrome

Marina Noris[1] *and Giuseppe Remuzzi*[1,2]

[1]Transplant Research Center "Chiara Cucchi de Alessandri e Gilberto Crespi", Mario Negri Institute for Pharmacological Research, Villa Camozzi, via Camozzi, 3, 24020 Ranica, Bergamo, Italy; [2]Department of Medicine and Transplantation, Ospedali Riuniti di Bergamo, Largo Barozzi 1, 24128, Bergamo, Italy

Definition

Hemolytic uremic syndrome (HUS) is a rare disease of microangiopathic hemolytic anemia, low platelet count and renal impairment [1]. Anemia is severe, non-immune (Coombs negative) and microangiopathic in nature, with fragmented red blood cells (schistocytes) in the peripheral smear, high serum lactate dehydrogenase (LDH), circulating free hemoglobin and reticulocytes. Platelet count is lower than 60,000/mm^3 in most cases [1], and platelet survival time is reduced, reflecting enhanced platelet disruption in the circulation.

In children, the disease is most commonly triggered by certain strains of *E. coli* that produce powerful exotoxins, the Shiga-like toxins (Stx-*E. coli*) [2–6], and manifests with diarrhea (D$^+$HUS), often bloody. Cases of Stx-*E. coli* HUS – around 25% [7] – that do not present with diarrhea have also been reported [8]. Acute renal failure manifests in 55–70% of cases [9–11]; however, renal function recovers in most of them (up to 70% in various series) [1, 8, 11, 12]. The overall incidence of Stx-HUS is estimated to be 2 cases per 100 000 persons/year, with a peak in summer months.

Non-Shiga toxin-associated HUS (non-Stx-HUS) comprises a heterogeneous group of patients in whom an infection by Stx-producing bacteria could be excluded as cause of the disease. It can be sporadic or familial (i.e., more than one member of a family affected by the disease and exposure to Stx-*E. coli* excluded). Collectively, non-Stx-HUS forms have a poor outcome. Up to 50% of cases progress to end-stage renal disease (ESRD) or have irreversible brain damage, and 25% may die during the acute phase of the disease [13–15].

Pathology

The common microvascular lesion of HUS, defined by the term thrombotic microangiopathy, consists of vessel wall thickening (mainly arterioles and capillar-

Complement and Kidney Disease, edited by Peter F. Zipfel

ies) with endothelial swelling and detachment, which allows accumulation of proteins and cell debris in the subendothelial layer, creating a space between endothelial cells and the underlying basement membrane of affected microvessels [1, 8] (Fig. 1). Both the widening of the subendothelial space and intraluminal platelet thrombi lead to a partial or complete obstruction of the vessel lumina. It is probably because of the partial occlusion of the lumen that erythrocytes are disrupted by mechanical trauma, which explains the Coombs-negative hemolysis and finding of fragmented and distorted erythrocytes in the blood smear.

In children younger than 2 years of age the lesion is mainly confined to the glomerular tuft and is noted in an early phase of the disease. Glomerular capillary lumina are reduced or occluded. In patent glomerular capillaries packed with red blood cells and fibrin, thrombi are occasionally seen. Examination of biopsy specimens taken several months after the disease onset showed that most glomeruli were normal (an indication of the reversibility of the lesions), whereas 20% eventually became sclerotic [16, 17]. Arterial thrombosis does occur but is uncommon, and appears to be a proximal extension of the glomerular lesion [16, 17]. In adults and older children, glomerular changes are different and more heterogeneous than in infants. The classical pattern of thrombotic microangiopathy is less evident. In these patients glomeruli mainly show ischemic changes with retraction of glomerular tufts, expansion of the urinary space, and thickening and wrinkling of the capillary wall (Fig. 2) [18]. When the renal biopsy is performed in the early phase of the disease, changes may be reminiscent of those of younger children with typical thickening of the glomerular capillary wall and swelling of the endothelium. Thrombi, packed and fragmented red blood cells are observed in the capillary lumina. Occasionally, glomeruli have necrotic areas or segmental extracapillary proliferation. Late in the course of the disease, thickening of the glomerular basement membrane and intraglomerular thrombi are less frequent and the ischemic changes prevail [17].

Mesangiolysis is observed in the early phase of the disease, characterized by enlargement of glomerular capillary loops with elongated and ectatic segments (Fig. 2). In subsequent phases, mesangiolysis is replaced by a form of nodular sclerosis called pale sclerosis, the nature of which remains to be established [17].

In the acute phase, tubular changes include foci of necrosis of proximal tubular cells, and the presence of red blood cells and eosinophilic casts in the lumina of distal tubules. Occasionally, fragmented red blood cells can be detected in the distal tubular lumina.

In children with a predominantly glomerular pattern, arteriolar lesions are rather heterogeneous: mild arteriolar lesions with some vessels not even affected by the microangiopathy were reported in a study [19], but arteriolar thrombosis was frequently detected in other childhood series [20, 21] with similar glomerular pattern. In adults, changes in the arterioles and arteries are common with severe narrowing of the lumina due to expansion of the subendothelial space. Fibrin and platelet thrombi and myointimal proliferation also are frequent (Figs 3, 4). On occasion,

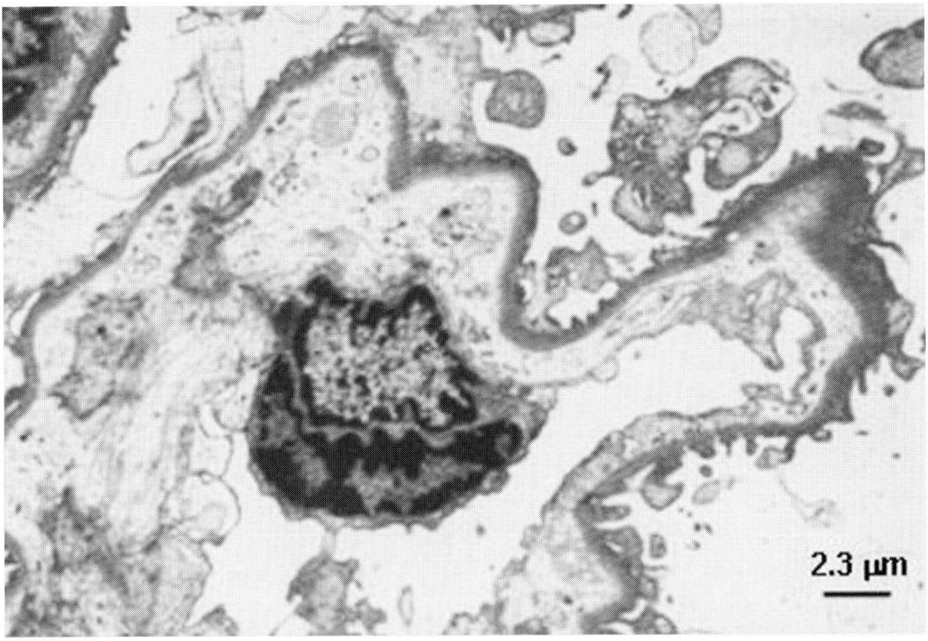

Figure 1
Electron micrograph of a glomerular capillary from a patient with HUS. The endothelium is detached from the glomerular basement membrane, the subendothelial space is widened and occupied by electron-lucent fluffy material and cell debris. The capillary lumen is markedly narrowed. Podocyte foot processes are focally effaced (magnification × 4400).

necrosis of the arteriolar wall can be detected. Arterial changes primarily affect the interlobular arteries. Vascular lumina often are congested with sludged and fragmented red blood cells. [16, 17]. Red blood cells and fibrin also may infiltrate the walls of arteries. Intimal proliferation has been regarded as a response to intimal swelling and to the accumulation of red blood cells and fibrin. When specimens are examined in the late phase of the disease, intimal thickening is replaced by dense intimal fibroplasias which leads to marked narrowing of the lumina.

At immunofluorescence, glomeruli almost invariably have fibrinogen deposits along the capillary wall and in the mesangium in a diffuse granular pattern. Staining is generally strong. Often fibrinogen antibodies stain large masses in glomerular capillaries that correspond to thrombi in the capillary lumina [19]. Arterioles and arteries are also positive for fibrinogen, which localizes both in the vessel wall and in thrombi. Granular deposits of IgM, C1q and C3 (see below) along the capillary loops of glomeruli were frequent in some series [19, 22]. Deposits of IgG and IgA were seldom detected [17].

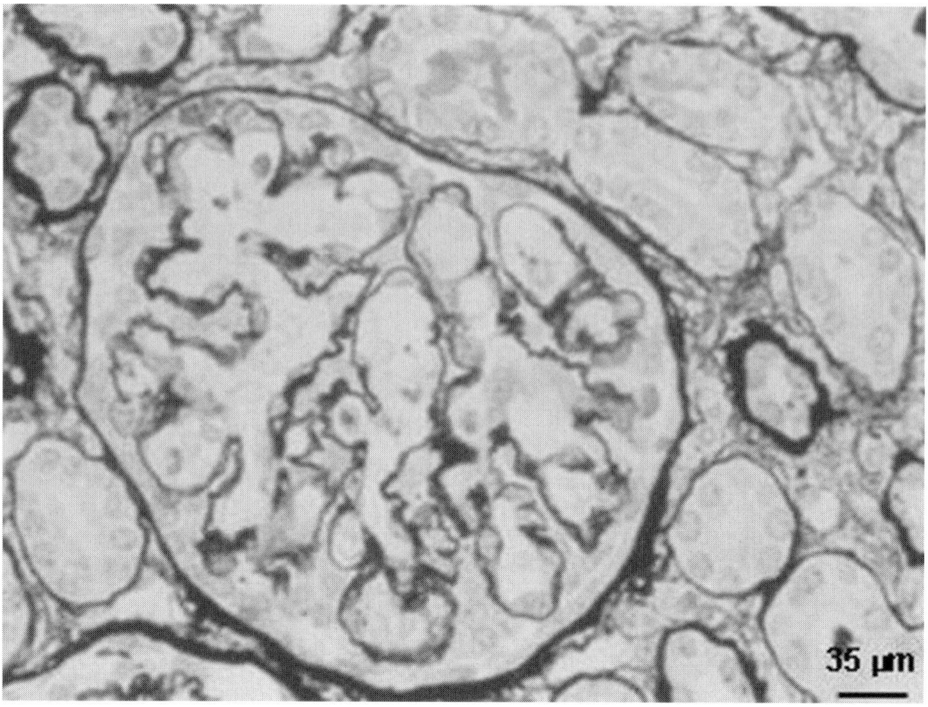

Figure 2
Glomerulus from a patient with HUS. A marked mesangiolysis (dissolution of mesangial matrix/anchor with dilated capillary lumina) and wrinkling of the capillary wall are evident (silver stain, magnification × 290).

Non-Stx-HUS

Epidemiology

Non-Stx-HUS is less common than Stx-HUS, and accounts for only 5–10% of all cases of the disease [1, 23]. It may manifest at all ages, but is more frequent in adults. According to a recent US study the incidence of non-Stx-HUS in children is approximately one tenth that of Stx-HUS [15], corresponding to approximately 2 cases per year per 1 000 000 total population. At variance with Stx-HUS, there is no clear etiologic agent or seasonal pattern. The onset may be preceded by features of the nephrotic syndrome. A diarrhea prodrome is rarely observed (diarrhea-negative HUS, D–HUS) [1, 8, 15, 24]. Non-Stx-HUS can occur sporadically or in families.

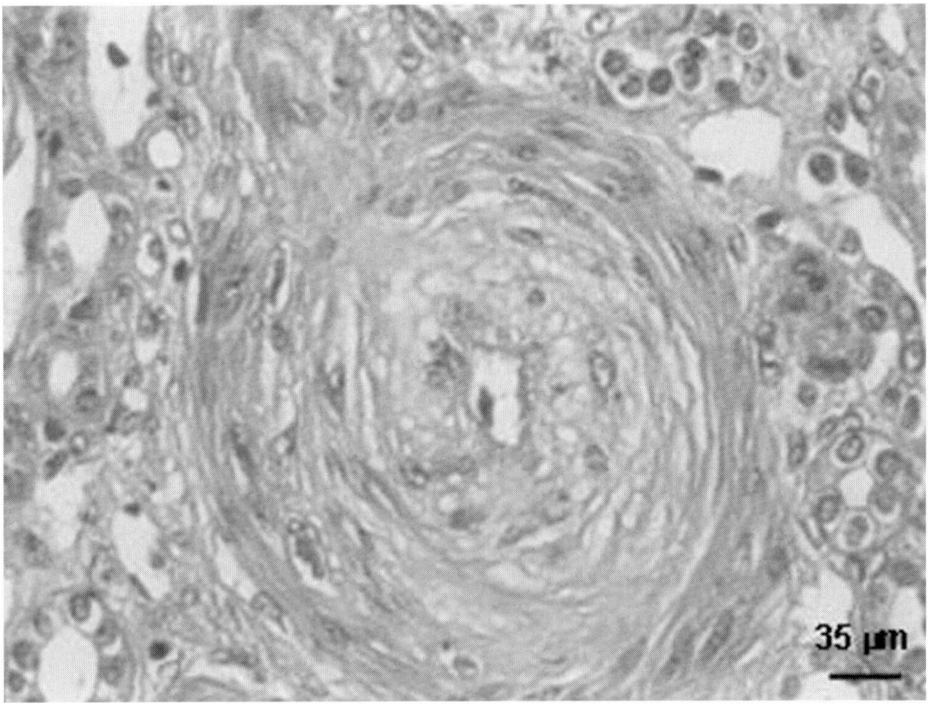

Figure 3
Small artery of a patient with HUS. Changes include mucoid intimal hyperplasia with severe narrowing of the lumen. (Periodic acid-Schiff; magnification × 290).

Sporadic non-Stx-HUS

A wide variety of triggers for sporadic non-Stx-HUS have been identified including various non-enteric infections, viruses, drugs, malignancies, transplantation, pregnancy and other underlying medical conditions (scleroderma, anti-phospholipid syndrome, lupus, malignant hypertension) (Tab. 1).

Infection caused by *Streptococcus pneumoniae* accounts for 40% of non-Stx-HUS and 4.7% of all causes of HUS in children in US [15]. Neuroaminidase produced by *Streptococcus pneumoniae*, by removing sialic acids from the cell membranes, exposes Thomsen-Friedenreich antigen to preformed circulating IgM antibodies, which bind to this neoantigen on platelets and endothelial cells and cause platelet aggregation and endothelial damage [25, 26]. The clinical picture is usually severe, with respiratory distress, neurological involvement and coma. The mortality rate is 50% [26].

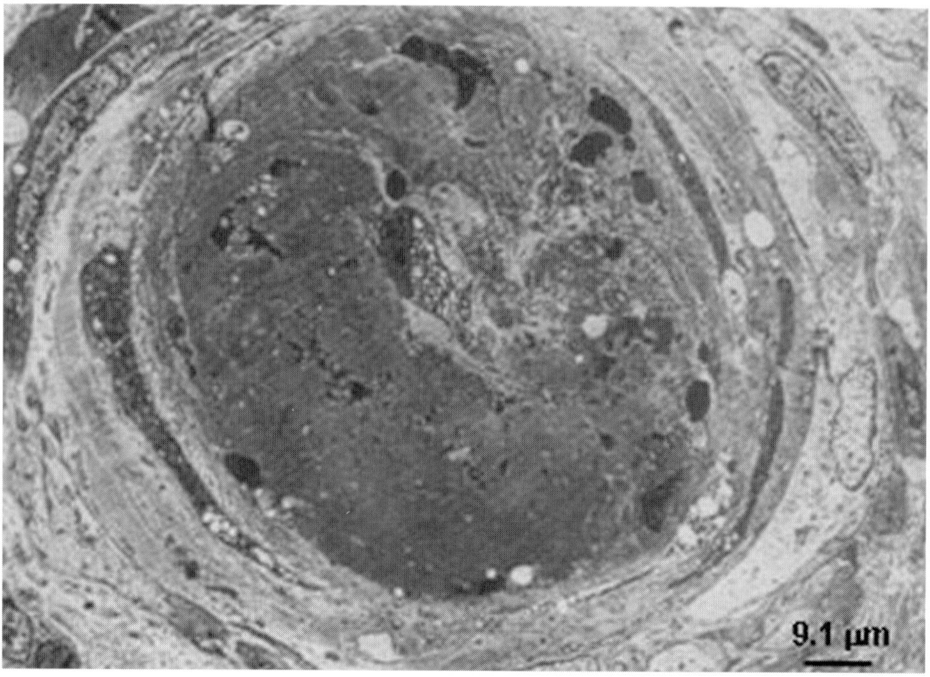

Figure 4
Electron micrograph of a renal arteriole in a case of HUS. The vascular lumen is completely occluded by amorphous material, red blood cell fragments and cell debris (magnification × 1100).

HUS and the related syndrome, thrombotic thrombocytopenic purpura (TTP) are both part of the complications of AIDS following HIV infection, and these forms account for up to 30% of hospitalized cases of HUS and TTP in some institutions [27]. The clinical course is poor and depends heavily on the severity of the underlying disease.

Categories of drugs that have been most frequently reported to induce non-Stx-HUS include anti-cancer molecules (mitomycin, cisplatin, bleomycin, gemcitabine), immunotherapeutic (cyclosporin, tacrolimus, OKT3, interferon and quinidine), and anti-platelet (ticlopidine, clopidogrel) agents [28]. The risk of developing HUS after mitomycin is 2–10%. The onset is delayed, occurring almost 1 year after starting treatment. The prognosis is poor, with up to 75% mortality at 4 months [28].

Post-transplant HUS is being reported with increasing frequency [1, 29]. It may ensue for the first time in patients who never suffered the disease (*de novo* post-transplant HUS) or may affect patients whose primary cause of ESRD was HUS (recurrent post-transplant HUS, discussed later in this chapter). *De novo* post-

Table 1 - Classification and treatment of different forms of Non-Stx-HUS

Disease	**Causes**	**Treatment**
Non-Stx-HUS		
Sporadic	Bacteria (*Streptococcus pneumoniae*)	Antibiotics, no plasma
	Viruses (HIV)	Plasma
	Drugs (antineoplastic, antiplatelet immunosuppressive)	Drug withdrawal, plasma
	Pregnancy-associated	Delivery, plasma
	Post-partum	Plasma
	Systemic diseases	
	- lupus	Steroids, plasma
	- scleroderma	BP control
	- anti-phospholipid syndrome	Oral anticoagulants
	Idiopathic	Plasma
	Genetic (factor H, MCP, factor I)	Plasma
Familial	Genetic (factor H, MCP, factor I)	Plasma

transplant HUS might occur in patients receiving renal transplants and other organs, as a consequence of the use of calcineurin inhibitors or due to humoral (C4b positive) rejection. In renal transplant patients treated with cyclosporin, the incidence of the disease is 5–15%; a lower incidence, approximately 1%, is found in patients receiving tacrolimus [30]. A peculiar form of *de novo* post-transplant HUS affects about 6% of the recipients of a bone marrow transplant (BMT), usually in the setting of a graft-versus-host disease (GVHD) or of intensive GVHD prophylaxis [30].

Non-Stx-HUS may also be triggered by pregnancy. Pregnancy-associated HUS may occasionally develop as a complication of pre-eclampsia. Some patients progress to a life-threatening variant of pre-eclampsia with severe thrombocytopenia, microangiopathic hemolytic anemia, renal failure and liver involvement (HELLP syndrome). HUS or its complications in pregnancy are always an indication for prompt delivery that is usually followed by complete remission [31]. Post-partum HUS is another complication of pregnancy that manifests within 3 months of delivery in most cases. The outcome is usually poor with 50–60% mortality; residual renal dysfunction and hypertension are the rule in surviving patients [32].

Of note, in about 50% of cases of sporadic non-Stx-HUS no clear triggering conditions could be found (idiopathic HUS) [1]. The outcome is variable, it may follow a progressive course to ESRD; however, many patients recover completely [1].

Familial non-Stx-HUS

Familial forms accounts for fewer than 3% of all cases of HUS. Reports of familial occurrence of HUS in children date back to 1965, when Campbell and Carre [33] described the development of hemolytic anemia and azotemia in concordant monozygous twins. Since then, familial forms of HUS in children and, less frequently, in adults, have been reported [34]. Although some reports have been in siblings, suggesting autosomal recessive transmission [35], there have also been others describing affected members across two [36–38] or three [35] successive generations, suggesting an autosomal dominant mode of inheritance. In a kindred [39], late onset in the grandmother and early onset in the grandson may be indicative of the phenomenon known as anticipation, in which clinical manifestations appear at an earlier age and with increasing severity in succeeding generations.

In autosomal recessive HUS, the onset is usually early in childhood. The prognosis is poor, with a mortality rate of 60–70%. Recurrences are very frequent.

The onset of symptoms in autosomal dominant HUS varies (early in childhood in some cases, but it can manifests in adult life), and is often triggered by precipitating events such as bacterial and viral infections or pregnancy [35]. The use of oral contraceptives and the intake of estrogens have also been implicated as triggering factors [35]. Malignant hypertension is a frequent complication and the prognosis is poor, with a cumulative incidence of death or ESRD of 50% [40, 41] to 90% [35]. In a large series of patients with familial history of HUS, identified through the database of the International Registry of Familial and Recurrent HUS/TTP, 54% had died because of the disease. Among survivors, 38% were on chronic dialysis [40]. Disease recurrences were reported in more than one half of the cases. Conditions predisposing to disease onset were recognized in 54% of cases, and included upper respiratory tract infections, pregnancy and consumption of birth control pills [40].

Complement abnormalities

Reduced serum levels of the third component (C3) of complement have been reported since 1974 in patients with both Stx-HUS and non-Stx-HUS, during the acute phase of the disease [37, 42]. In patients suffering from episodes of Stx-HUS C3 abnormalities consistently subside with remission of the disease. In contrast, in cases of familial non-Stx-HUS and in patients suffering from recurrences, serum C3 levels were consistently and remarkably low, even during remission of the disease [40, 43, 44]. However, levels of the C4 are usually normal. Persistently low C3 levels in patients with familial HUS, even in the unaffected relatives, suggested a genetic defect [40]. Low C3 levels most likely reflect complement activation and C3 consumption, which is consistent with increased levels of C3 activation and breakdown products, C3b, C3c and C3d, in the plasma of these patients [45–47]. Granular C3 deposits in glomeruli and arterioles of kidney biopsies taken during the acute phase of the disease [22, 48–51] further reflect activation of the complement system and C3 consumption in the renal microvasculature.

The complement system consists of several plasma and membrane-associated proteins that are organized in three distinct activation patterns. These include classical, lectin and alternative pathways [52, 53] that, once activated on the surface of microorganisms, form protease complexes, collectively known as C3 convertases, which serve to cleave C3 generating C3b. The classic/lectin convertases are formed by C2 and C4 fragments, while the generation of the alternative pathway convertase requires the cleavage of C3 and Factor B, but not of C4. Thus, low C3 levels in patients with HUS in the presence of normal C4 indicate a selective activation of the alternative pathway. Reduction of Factor B levels and, particularly, the appearance of Factor B (Ba and Bb) breakdown products have been also reported in HUS patients [46], confirming the activation of the alternative pathway of complement. Upon generation C3b deposits on bacterial surfaces, which leads to opsonization for phagocytosis by neutrophils and macrophages. C3b also participates in the formation of the C5 convertases that cleaves C5 and initiates assembly of the membrane attack complex (MAC), causing cell lysis. The human complement system is highly regulated to prevent nonspecific damage to host cells and to limit deposition of C3b to the surface of pathogens. This fine regulation is based on a number of membrane-anchored (CR1, DAF, MCP and CD59) and fluid-phase (Factor H and Factor I) regulators that protect host tissues. Foreign surfaces that either lack membrane-bound regulators or cannot bind soluble regulators are attacked by complement.

Evidence is now emerging that the activation of the alternative pathway of complement in familial forms of non-Stx-HUS may derive from genetic abnormalities of some of the molecules, mentioned above, that are involved in the regulation of the complement system. Similar genetic abnormalities have been found in sporadic non-Stx-HUS as well. Thus, more than 50 different Factor H mutations have been identified in 80 patients who had familial and sporadic forms of HUS [38, 41, 54–59]. The Factor H mutation frequency is up to 40% in familial forms and 13–17% in sporadic forms. In addition, mutations for the cell-bound membrane complement regulator, membrane cofactor protein (MCP), have been reported in familial and sporadic non-Stx-HUS, accounting for around 10% mutation frequency [45, 60]. Finally, three mutations in the gene encoding for Factor I have been reported in three patients with the sporadic form [61], which supports the idea that non-Stx-HUS is a disease of complement dysregulation.

Complement regulatory proteins are essential to prevent nonspecific damage of host endothelial cells and limit deposition of C3b to the surface of pathogens [52, 53]. Thus, in patients with genetic alterations causing defective activity of complement regulatory proteins, exposure to an agent that activates complement results in C3b deposition on vascular endothelial cells and in the formation of the C3 and C5 convertase complexes. The proteolysis of C3 and C5 by convertases causes the release of chemotactic anaphylatoxins (C3a and C5a) that recruit inflammatory cells toward the endothelial layer. On the other hand, the deposition of C3b is followed by the formation of MAC, which leads to endothelial cells injury, with increased expression of adhesion molecules and tissue factor, and endothelial detachment. Under these conditions, platelets from the microcirculation adhere and aggregate, and the formation of thrombin and of fibrin polymers occurs.

Of note, not all patients with non-Stx-HUS have lower than normal C3 serum concentrations, even in the presence of genetic alterations of complement regulators. Thus, the incidence of hypocomplementemia ranges between 31% and more than 90% in patients with Factor H mutations and between 0% and 51% in patients with no Factor H mutations, according to different series [41, 54–56, 62, 63]. Similarly, 33–50% of patients with MCP mutations [45, 60] and two-thirds of patients with Factor I mutations [61] have lower than normal serum C3 levels. Thus, normal C3 levels in patients with non-Stx-HUS do not necessarily exclude a genetic deficiency of complement regulation. More sensitive markers of complements activation status are increased plasma levels of C3 breakdown products and an increased C3d/C3 ratio, which have been found altered in most patients with HUS even in the face of normal serum C3 concentrations [45–47]. Another consistent marker of complement activation could be the presence of C3 deposits in renal biopsies. Significant amounts of C3 deposits were found in the glomeruli of renal biopsy specimens from eight children with renal sequelae following a severe HUS episode, despite serum C3 levels being normal [49]. C3 deposits appeared as nodular pattern along the capillary walls. Immunohistology evaluation of renal biopsy from two siblings with familial HUS and normal C3 serum levels evidenced the presence of C3 deposits in renal arterioles and arteries [64]. In another report, examination of kidney biopsy material from two siblings of a big Bedouin family with ten affected infants due to an homozygous Factor H mutation, revealed strong C3 and MAC formation and C5b-9 deposition in glomerular capillary walls [65]. Similar findings were reported in the kidney biopsy material from another familial case with an adult onset and an heterozygous Factor H mutation [41] (Fig. 5).

Treatment of non-Stx-HUS

Despite the fact that non-Stx-HUS has a rather poor prognosis, after plasma manipulation was introduced, the mortality rate has dropped from 50% to 25% [66–68] (Tab. 1). However, debate still exists on whether plasma is or is not effective in the treatment of acute episodes [69–72]. Published observations [67, 73–75] and our own experience indicate that a consistent number of patients with non-Stx-HUS respond to plasma treatment. It has been proposed that plasma exchange might be relatively more effective than plasma infusion since it might remove potentially toxic substances from the patient's circulation (see chapter by Würzner & Zimmerhackl).

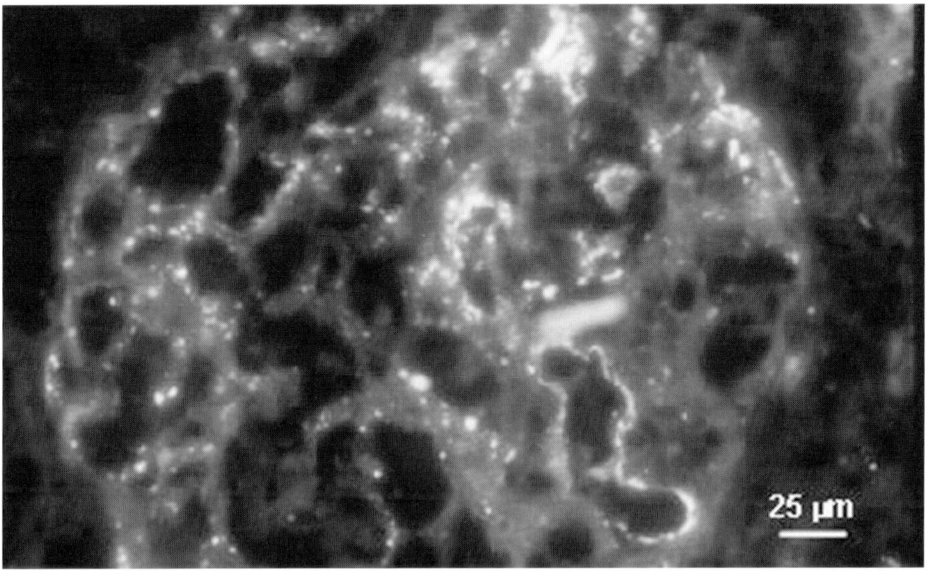

Figure 5
Glomerulus from a patient with HUS carrying a heterozygous mutation (delA1494-1496, causing a premature interruption of the protein at CCP 8) in Factor H gene, stained with a fluorescent anti-C3 antibody. Strong staining of capillaries, and many subendothelial granular deposits are evident (magnification × 400).

That this may not be the case is documented by data that, in a patient with relapsing thrombotic microangiopathy [76], normalization of the platelet count was invariably obtained by plasma exchange or infusion, whereas plasma removal and substitution with albumin and saline never raised the platelet count. However, in situations such as renal insufficiency or heart failure that limit the amount of plasma that can be provided with infusion alone, plasma exchange should be considered as first-choice therapy [1]. Plasma infusion or exchange has been used in patients with HUS and Factor H mutations, with the rationale to provide the patients with normal Factor H to correct the genetic deficiency. Some patients did not respond at all and died or developed ESRD [65]. Others remained chronically ill [77, 78] or required infusion of plasma at weekly intervals to rise Factor H plasma levels enough to maintain remission [79]. Stratton et al. [80] were able to induce sustained remission in a patient with Factor H mutation who developed an acute episode of HUS and required hemodialysis. After 3 months of weekly plasma exchange in conjunction with intravenous immunoglobulins, the patient regained renal function, dialysis was withdrawal and plasma therapy stopped. At 1 year after stopping plasma therapy, the patient remained disease-free and dialysis independent.

Plasma therapy is contraindicated in patients with HUS induced by *Streptococcus pneumoniae*, because adult plasma contains antibodies against the Thomsen-Friedenreich antigen, which may exacerbate the disease (Tab. 1).

In those few patients with extensive microvascular thrombosis at renal biopsy, refractory hypertension and signs of hypertensive encephalopathy, and for whom conventional therapies including plasma manipulation are not enough to control the disease (i.e., persistent severe thrombocytopenia and hemolytic anemia), bilateral nephrectomy has been performed with excellent follow-up in some patients [81]. Other treatments including anti-platelet agents, prostacyclin, heparin or fibrinolytic agents, steroids and intravenous immunoglobulins have been attempted, with no consistent benefit [1].

Those patients who develop HUS upon challenge with cyclosporin or tacrolimus have to stop the medication. Sirolimus has been used as an alternative in occasional patients with encouraging results [82].

Kidney transplantation in non-Stx-HUS

Less than 15% of children with Stx-HUS progress to ESRD, but when it happens renal transplantation can be safely performed. Thus, in children with Stx-HUS disease recurrence on the graft ranges from 0% to 10% [83, 84], and graft survival at 10 years is even better than in control transplanted children with other diseases [85]. On the other hand, 50% (in sporadic forms) to 60% (in familial forms) of patients with non-Stx-HUS [1, 41] progress to ESRD. Renal transplantation is not necessarily an option for non-Stx-HUS, and whether kidney transplantation is feasible in these patients is still a debated issue. Actually, recurrence occurred in over 50% of the patients who had a renal transplant, with graft failure developing in around 90% of them despite any treatment. [84, 86]. Recurrences occur at a median time of 30 days after transplant (range 0 days–16 years). [1, 84]. Neither avoidance of calcineurin inhibitors as part of the immunosuppressive therapy nor bilateral nephrectomy before transplantation prevented recurrence of HUS and graft loss. Patients who lost the first kidney graft for recurrence should not be re-transplanted, since the disease recurs in around 90% of the subsequent grafts. Live-related renal transplant should also be avoided, in that it carries the additional risk to precipitate the disease onset in the healthy donor relative, as recently reported in two families [87]. New knowledge from genetic studies would hopefully allow the risk of recurrence to be more accurately predicted. In patients with Factor H mutations, the recurrence rate ranges from 30% to 100% according to different surveys [41, 54, 55], and is significantly higher than in patients without Factor H mutations [41]. In view of the fact that Factor H is a plasma protein mainly of liver origin, a kidney transplant does not correct the Factor H genetic defect [38, 58].

Simultaneous kidney and liver transplant has been recently performed in two young children with non-Stx-HUS and Factor H mutations, with the objective of correcting the genetic defect and prevent disease recurrences [88, 89]. However, for reasons that are currently under evaluation and that possibly involve an increased liver susceptibility to ischemic or immune injury related to uncontrolled complement activation, both cases treated with this procedure were complicated by premature irreversible liver failure. In the first patient, a humoral rejection of the liver graft manifested by the day 26 after transplantation and in a few days the child developed hepatic encephalopathy and coma, and recovered with a second, uneventful, liver transplant [88]. The second case was complicated by a fatal, primary nonfunction of the liver graft followed by multi-organ failure and patient's death [89]. Thus, despite its capacity of correcting the genetic defect, combined kidney and liver transplant for non-Stx-HUS associated with Factor H mutations, should not be performed unless a patient is at imminent risk of life-threatening complications.

On the other hand, 25–50% of patients with non-Stx-HUS but no evidence of Factor H mutations, who received a kidney transplant, lost the graft within 1 year [41, 54]. A remarkable exception are possibly those forms associated with MCP mutations. Four of these patients have been transplanted successfully with no disease recurrence [60] (and our unpublished data). There is a strong biochemical rationale for that: at variance with Factor H, which is a circulating soluble inhibitor, MCP is a membrane-bound protein highly expressed in the kidney. The transplantation of a normal kidney in the latter condition not surprisingly corrects the genetic defect. The case of Factor I, again a soluble and circulating inhibitor of liver origin, should not be different from the case of Factor H. Actually, the occurrence of two graft failures in a recently published case with a Factor I mutation [61] seems to be consistent with this reasoning.

Thus, genetic testing should be performed in all patients with ESRD due to non-Stx-HUS awaiting transplantation. Renal transplantation is strongly contraindicated in those patients with Factor H mutations due to the high risk of disease recurrence. At variance, graft outcome is good in patients with mutations in MCP gene, and renal transplantation should be recommended in these patients.

Despite recent advances, the underlying genetic alteration remains unknown in more than one half of non-Stx-HUS patients, and further studies are required to fully clarify the possible role of additional candidate genes encoding for proteins involved in the regulation of complement, which would hopefully allow in the future a more tailored clinical management of these patients.

Acknowledgements

This work has been partially supported by grants from Comitato 30 ore per la vita, from Telethon (grants GPP02161 and GPP02162) and from Foundation for children with Atypical HUS along with The Nando Peretti Foundation.

References

1 Ruggenenti P, Noris M, Remuzzi G (2001) Thrombotic microangiopathy, hemolytic uremic syndrome, and thrombotic thrombocytopenic purpura. *Kidney Int* 60: 831–846

2 Thorpe CM (2004) Shiga toxin-producing *Escherichia coli* infection. *Clin Infect Dis* 38: 1298–1303

3 Karmali MA, Steele BT, Petric M, Lim C (1983) Sporadic cases of haemolytic-uraemic syndrome associated with faecal cytotoxin and cytotoxin-producing *Escherichia coli* in stools. Lancet 1: 619–620

4 Riley LW, Remis RS, Helgerson SD, McGee HB, Wells JG, Davis BR, Hebert RJ, Olcott ES, Johnson LM, Hargrett NT, Blake PA, Cohen ML (1983) Hemorrhagic colitis associated with a rare *Escherichia coli* serotype. *N Engl J Med* 308: 681–685

5 Mead PS, Griffin PM (1998) *Escherichia coli* O157:H7. Lancet 352: 1207–1212

6 Jackson MP, Newland JW, Holmes RK, O'Brien AD (1987) Nucleotide sequence analysis of the structural genes for Shiga-like toxin I encoded by bacteriophage 933J from *Escherichia coli. Microb Pathog* 2: 147–153

7 Gianviti A, Rosmini F, Caprioli A, Corona R, Matteucci MC, Principato F, Luzzi I, Rizzoni G (1994) Haemolytic-uraemic syndrome in childhood: surveillance and case-control studies in Italy. Italian HUS Study Group. *Pediatr Nephrol* 8: 705–709

8 Kaplan BS, Meyers KE, Schulman SL (1998)The pathogenesis and treatment of hemolytic uremic syndrome. *J Am Soc Nephrol* 9: 1126–1133

9 Tonshoff B, Sammet A, Sanden I, Mehls O, Waldherr R, Scharer K (1994) Outcome and prognostic determinants in the hemolytic uremic syndrome of children. *Nephron* 68: 63–70

10 Banatvala N, Griffin PM, Greene KD, Barrett TJ, Bibb WF, Green JH, Wells JG (2001)The United States National Prospective Hemolytic Uremic Syndrome Study: microbiologic, serologic, clinical, and epidemiologic findings. *J Infect Dis* 183: 1063–1070

11 Milford D (1992) The hemolytic uremic syndromes in the United Kingdom. In: BSTR Kaplan, JL Moake (eds): *Hemolytic Uremic Syndrome and Thrombotic Thrombocytopenic Purpura.* Marcel Dekker Inc., New York, 39–59

12 Van Dyck M, Proesmans W, Depraetere M (1988) Hemolytic uremic syndrome in childhood: renal function ten years later. *Clin Nephrol* 29: 109–112

13 Schieppati A, Ruggenenti P, Cornejo RP, Ferrario F, Gregorini G, Zucchelli P, Rossi E, Remuzzi G (1992) Renal function at hospital admission as a prognostic factor in adult hemolytic uremic syndrome. The Italian Registry of Haemolytic Uremic Syndrome. *J Am Soc Nephrol* 2: 1640–1644

14 Taylor CM, Chua C, Howie AJ, Risdon RA (2004) Clinico-pathological findings in diarrhoea-negative haemolytic uraemic syndrome. *Pediatr Nephrol* 19: 419–425

15 Constantinescu AR, Bitzan M, Weiss LS, Christen E, Kaplan BS, Cnaan A, Trachtman H (2004) Non-enteropathic hemolytic uremic syndrome: causes and short-term course. *Am J Kidney Dis* 43: 976–982

16 Taylor CM, Howie AJ, Williams JM (1999) No common final pathogenetic pathway in haemolytic uraemic syndromes. *Nephrol Dial Transplant* 14: 1100–1102
17 Remuzzi G, Ruggenenti P (1994) Thrombotic microangiopathies. In: C Tisher, B Brenner (eds): *Renal Pathology*, 2nd ed. J.B. Lippincott, Philadelphia, 1154–1184
18 Kincaid-Smith P (1975) Participation of intravascular coagulation in the pathogenesis of glomerular and vascular lesions. *Kidney Int* 7: 242–253
19 Levy M, Gagnadoux M, Habib R (1980) Pathology of hemolytic uremic syndrome in children. In: G Remuzzi, G Mecca, G dG (eds): *Hemostasis, Prostaglandins and Renal Disease*. Raven Press, New York, 383–397
20 Vitsky BH, Suzuki Y, Strauss L, Churg J(1969) The hemolytic-uremic syndrome: a study of renal pathologic alterations. *Am J Pathol* 57: 627–647
21 Proia AD, Harden EA, Silberman HR (1984) Mitomycin-induced hemolytic-uremic syndrome. *Arch Pathol Lab Med* 108: 959–962
22 Hammar SP, Bloomer HA, McCloskey D (1978) Adult hemolytic uremic syndrome with renal arteriolar deposition of IgM andC3. *Am J Clin Pathol* 70: 434–439
23 Hall SM, Glickman M (1988) The British Paediatric Surveillance Unit. *Arch Dis Child* 63: 344–346
24 Renaud C, Niaudet P, Gagnadoux MF, Broyer M, Habib R (1995) Haemolytic uraemic syndrome: prognostic factors in children over 3 years of age. *Pediatr Nephrol* 9: 24–29
25 Erickson LC, Smith WS, Biswas AK, Camarca MA, Waecker NJ Jr (1994) Streptococcus pneumoniae-induced hemolytic uremic syndrome: a case for early diagnosis. *Pediatr Nephrol* 8: 211–213
26 Novak RW, Martin CR, Orsini EN (1983) Hemolytic-uremic syndrome and T-cryptantigen exposure by neuraminidase-producing pneumococci: an emerging problem? *Pediatr Pathol* 1: 409–413
27 Leaf AN, Laubenstein LJ, Raphael B, Hochster H, Baez L, Karpatkin S (1988)Thrombotic thrombocytopenic purpura associated with human immunodeficiency virus type 1 (HIV-1) infection. *Ann Intern Med* 109: 194–197
28 Dlott JS, Danielson CF, Blue-Hnidy DE, McCarthy LJ (2004) Drug-induced thrombotic thrombocytopenic purpura/hemolytic uremic syndrome: a concise review. *Therap Apher Dial* 8: 102–111
29 Reynolds JC, Agodoa LY, Yuan CM, Abbott KC (2003) Thrombotic microangiopathy after renal transplantation in the United States. *Am J Kidney Dis* 42: 1058–1068
30 Ruggenenti P (2002) Post-transplant hemolytic-uremic syndrome. *Kidney Int* 62: 1093–1104
31 Remuzzi G, Ruggenenti P (1995)The hemolytic uremic syndrome. *Kidney Int* 48: 2–19
32 George JN (2003) The association of pregnancy with thrombotic thrombocytopenic purpura-hemolytic uremic syndrome. *Curr Opin Hematol* 10: 339–344
33 Campbell S, Carre IJ (1965)Fatal haemolytic uraemic syndrome and idiopathic hyperlipaemia in monozygotic twins. *Arch Dis Child* 40: 654–658
34 Kaplan BS, Chesney RW, Drummond KN (1975) Hemolytic uremic syndrome in families. *N Engl J Med* 292: 1090–1093

35 Berns JS, Kaplan BS, Mackow RC, Hefter LG (1992) Inherited hemolytic uremic syndrome in adults. *Am J Kidney Dis* 19: 331–334
36 Noris M, Azzollini N, Mister M, Pezzotta A, Piccinini G, Casiraghi F, Cugini D, Perico N, Orisio S, Remuzzi G (1999) Peripheral donor leukocytes prolong survival of rat renal allografts. *Kidney Int* 56: 1101–1112
37 Carreras L, Romero R, Requesens C, Oliver AJ, Carrera M, Clavo M, Alsina J (1981) Familial hypocomplementemic hemolytic uremic syndrome with HLA-A3,B7 haplotype. *Jama* 245: 602–604
38 Warwicker P, Goodship TH, Donne RL, Pirson Y, Nicholls A, Ward RM, Turnpenny P, Goodship JA (1988) Genetic studies into inherited and sporadic hemolytic uremic syndrome. *Kidney Int* 53: 836–844
39 Petermann A, Offermann G, Distler A, Sharma AM (1998) Familial hemolytic-uremic syndrome in three generations. *Am J Kidney Dis* 32: 1063–1067
40 Noris M, Ruggenenti P, Perna A, Orisio S, Caprioli J, Skerka C, Vasile B, Zipfel PF, Remuzzi G (1999) Hypocomplementemia discloses genetic predisposition to hemolytic uremic syndrome and thrombotic thrombocytopenic purpura: role of factor H abnormalities. Italian Registry of Familial and Recurrent Hemolytic Uremic Syndrome/ Thrombotic Thrombocytopenic Purpura. *J Am Soc Nephrol* 10: 281–293
41 Caprioli J, Castelletti F, Bucchioni S, Bettinaglio P, Bresin E, Pianetti G, Gamba S, Brioschi S, Daina E, Remuzzi G, Noris M (2003) Complement factor H mutations and gene polymorphisms in haemolytic uraemic syndrome: the C-257T, the A2089G and the G2881T polymorphisms are strongly associated with the disease. *Hum Mol Genet* 12: 3385–3395
42 Stuhlinger W, Kourilsky O, Kanfer A, Sraer JD (1974) Letter: Haemolytic-uraemic syndrome: evidence for intravascular C3 activation. *Lancet* 2: 788–789
43 Bogdanovic R, Cvoric A, Nikolic V, Sindjic M (1988) Recurrent haemolytic-uraemic syndrome with hypocomplementaemia: a case report. *Pediatr Nephrol* 2: 236–238
44 Pichette V, Querin S, Schurch W, Brun G, Lehner-Netsch G, Delage JM (1994) Familial hemolytic-uremic syndrome and homozygous factor H deficiency. *Am J Kidney Dis* 24: 936–941
45 Noris M, Brioschi S, Caprioli J, Todeschini M, Bresin E, Porrati F, Gamba S, Remuzzi G (2003) Familial haemolytic uraemic syndrome and an MCP mutation. *Lancet* 362: 1542–1547
46 Kim Y, Miller K, Michael AF (1977) Breakdown products of C3 and factor B in hemolytic-uremic syndrome. *J Lab Clin Med* 89: 845–850
47 Monnens L, Molenaar J, Lambert PH, Proesmans W, van Munster P (1980) The complement system in hemolytic-uremic syndrome in childhood. *Clin Nephrol* 13: 168–171
48 Gonzalo A, Mampaso F, Gallego N, Bellas C, Segui J, Ortuno J (1981) Hemolytic uremic syndrome with hypocomplementemia and deposits of IgM and C3 in the involved renal tissue. *Clin Nephrol* 16: 193–199
49 Cossio PM, Laguens RP, Pantin DJ, de Bracco MM, Molinas F, Voyer LE, Arana RM

(1977) Persistent glomerulonephritis following the haemolytic-uremic syndrome. Immunopathological and morphological studies. *Clin Exp Immunol* 29: 361–368
50 Barre P, Kaplan BS, de Chadarevian JP, Drummond KN (1977) Hemolytic uremic syndrome with hypocomplemententemia, serum C3NeF, and glomerular deposits of C3. *Arch Pathol Lab Med* 101: 357–361
51 Ohali M, Shalev H, Schlesinger M, Katz Y, Kachko L, Carmi R, Sofer S, Landau D (1998) Hypocomplemententemic autosomal recessive hemolytic uremic syndrome with decreased factor H. *Pediatr Nephrol* 12: 619–624
52 Walport MJ (2001) Complement. First of two parts. *N Engl J Med* 344: 1058–1066
53 Walport MJ(2001) Complement. Second of two parts. *N Engl J Med* 344: 1140–1144
54 Neumann HP, Salzmann M, Bohnert-Iwan B, Mannuelian T, Skerka C, Lenk D, Bender BU, Cybulla M, Riegler P, Konigsrainer A et al. (2003) Haemolytic uraemic syndrome and mutations of the factor H gene: a registry-based study of German speaking countries. *J Med Genet* 40: 676–681
55 Dragon-Durey MA, Fremeaux-Bacchi V, Loirat C, Blouin J, Niaudet P, Deschenes G, Coppo P, Herman Fridman W, Weiss L (2004) Heterozygous and homozygous factor h deficiencies associated with hemolytic uremic syndrome or membranoproliferative glomerulonephritis: report and genetic analysis of 16 cases. *J Am Soc Nephrol* 15: 787–795
56 Perez-Caballero D, Gonzalez-Rubio C, Gallardo ME, Vera M, Lopez-Trascasa M, Rodriguez de Cordoba S, Sanchez-Corral P(2001) Clustering of missense mutations in the C-terminal region of factor H in atypical hemolytic uremic syndrome. *Am J Hum Genet* 68: 478-484
57 Rodriguez de Cordoba S, Esparza-Gordillo J, Goicoechea de Jorge E, Lopez-Trascasa M, Sanchez-Corral P (2004) The human complement factor H: functional roles, genetic variations and disease associations. *Mol Immunol* 41: 355–367
58 Caprioli J, Bettinaglio P, Zipfel PF, Amadei B, Daina E, Gamba S, Skerka C, Marziliano N, Remuzzi G, Noris M (2001) The molecular basis of familial hemolytic uremic syndrome: mutation analysis of factor H gene reveals a hot spot in short consensus repeat 20. *J Am Soc Nephrol* 12: 297–307
59 Buddles MR, Donne RL, Richards A, Goodship J, Goodship TH (2000) Complement factor H gene mutation associated with autosomal recessive atypical hemolytic uremic syndrome. *Am J Hum Genet* 66: 1721–1722
60 Richards A, Kemp EJ, Liszewski MK, Goodship JA, Lampe AK, Decorte R, Muslumanoglu MH, Kavukcu S, Filler G, Pirson Y et al. (2003) Mutations in human complement regulator, membrane cofactor protein (CD46), predispose to development of familial hemolytic uremic syndrome. *Proc Natl Acad Sci USA* 100: 12966–12971
61 Fremeaux-Bacchi V, Dragon-Durey MA, Blouin J, Vigneau C, Kuypers D, Boudailliez B, Loirat C, Rondeau E, Fridman WH (2004) Complement factor I: a susceptibility gene for atypical haemolytic uraemic syndrome. *J Med Genet* 41: e84
62 Richards A, Buddles MR, Donne RL, Kaplan BS, Kirk E, Venning MC, Tielemans CL, Goodship JA, Goodship TH (2001) Factor H mutations in hemolytic uremic syndrome

cluster in exons 18-20, a domain important for host cell recognition. *Am J Hum Genet* 68: 485–490

63 Rougier N, Kazatchkine MD, Rougier JP, Fremeaux-Bacchi V, Blouin J, Deschenes G, Soto B, Baudouin V, Pautard B, Proesmans W et al (1998) Human complement factor H deficiency associated with hemolytic uremic syndrome. *J Am Soc Nephrol* 9: 2318–2326

64 Pirson Y, Lefebvre C, Arnout C, van Ypersele de Strihou C (1987) Hemolytic uremic syndrome in three adult siblings: a familial study and evolution. *Clin Nephrol* 28: 250–255

65 Landau D, Shalev H, Levy-Finer G, Polonsky A, Segev Y, Katchko L (2001) Familial hemolytic uremic syndrome associated with complement factor H deficiency. *J Pediatr* 138: 412–417

66 Hollenbeck M, Kutkuhn B, Aul C, Leschke M, Willers R, Grabensee B (1998) Haemolytic-uraemic syndrome and thrombotic-thrombocytopenic purpura in adults: clinical findings and prognostic factors for death and end-stage renal disease. *Nephrol Dial Transplant* 13: 76–81

67 Lara PN Jr, Coe TL, Zhou H, Fernando L, Holland PV, Wun T(1999) Improved survival with plasma exchange in patients with thrombotic thrombocytopenic purpura-hemolytic uremic syndrome. *Am J Med* 107: 573–579

68 Morel-Maroger L, Kanfer A, Solez K, Sraer JD, Richet G (1979)Prognostic importance of vascular lesions in acute renal failure with microangiopathic hemolytic anemia (hemolytic-uremic syndrome): clinicopathologic study in 20 adults. *Kidney Int* 15: 548–558

69 Dundas S, Murphy J, Soutar RL, Jones GA, Hutchinson SJ, Todd WT (1999) Effectiveness of therapeutic plasma exchange in the 1996 Lanarkshire *Escherichia coli* O157:H7 outbreak. *Lancet* 354: 1327–1330

70 Tsai HM (2003) Is severe deficiency of ADAMTS-13 specific for thrombotic thrombocytopenic purpura? Yes. *J Thromb Haemost* 1: 625–631

71 Tsai HM (2003) Deficiency of ADAMTS-13 in thrombotic and thrombocytopenic purpura. *J Thromb Haemost* 1: 2038–2040; discussion 2040–2035

72 Remuzzi G (2003): Is ADAMTS-13 deficiency specific for thrombotic thrombocytopenic purpura? No. *J Thromb Haemost* 1: 632–634

73 Clark WF, Rock GA, Buskard N, Shumak KH, LeBlond P, Anderson D, Sutton DM (1999)Therapeutic plasma exchange: an update from the Canadian Apheresis Group. *Ann Intern Med* 131: 453–462

74 Vesely SK, George JN, Lammle B, Studt JD, Alberio L, El-Harake MA, Raskob GE (2003) ADAMTS13 activity in thrombotic thrombocytopenic purpura-hemolytic uremic syndrome: relation to presenting features and clinical outcomes in a prospective cohort of 142 patients. *Blood* 102: 60–68

75 Tostivint I, Mougenot B, Flahault A, Vigneau C, Costa MA, Haymann JP, Sraer JD, Rondeau E (2002) Adult haemolytic and uraemic syndrome: causes and prognostic factors in the last decade. *Nephrol Dial Transplant* 17: 1228–1234

76 Ruggenenti P, Galbusera M, Cornejo RP, Bellavita P, Remuzzi G (1993) Thrombotic thrombocytopenic purpura: evidence that infusion rather than removal of plasma induces remission of the disease. *Am J Kidney Dis* 21: 314–318

77 Filler G, Radhakrishnan S, Strain L, Hill A, Knoll G, Goodship TH (2004) Challenges in the management of infantile factor H associated hemolytic uremic syndrome. *Pediatr Nephrol* 19: 908–911

78 Gerber A, Kirchhoff-Moradpour AH, Obieglo S, Brandis M, Kirschfink M, Zipfel PF, Goodship JA, Zimmerhackl LB (2003) Successful (?) therapy of hemolytic-uremic syndrome with factor H abnormality. *Pediatr Nephrol* 18: 952–955

79 Nathanson S, Fremeaux-Bacchi V, Deschenes G (2001) Successful plasma therapy in hemolytic uremic syndrome with factor H deficiency. *Pediatr Nephrol* 16: 554–556

80 Stratton JD, Warwicker P (2002) Successful treatment of factor H-related haemolytic uraemic syndrome. *Nephrol Dial Transplant* 17: 684–685

81 Remuzzi G, Galbusera M, Salvadori M, Rizzoni G, Paris S, Ruggenenti P (1996) Bilateral nephrectomy stopped disease progression in plasma-resistant hemolytic uremic syndrome with neurological signs and coma. *Kidney Int* 49: 282–286

82 Franco A, Hernandez D, Capdevilla L, Errasti P, Gonzalez M, Ruiz JC, Sanchez J (2003) *De novo* hemolytic-uremic syndrome/thrombotic microangiopathy in renal transplant patients receiving calcineurin inhibitors: role of sirolimus. *Transplant Proc* 35: 1764–1766

83 Loirat C, Niaudet P (2003) The risk of recurrence of hemolytic uremic syndrome after renal transplantation in children. *Pediatr Nephrol* 18: 1095–1101

84 Artz MA, Steenbergen EJ, Hoitsma AJ, Monnens LA, Wetzels JF (2003) Renal transplantation in patients with hemolytic uremic syndrome: high rate of recurrence and increased incidence of acute rejections. *Transplantation* 76: 821–826

85 Ferraris JR, Ramirez JA, Ruiz S, Caletti MG, Vallejo G, Piantanida JJ, Araujo JL, Sojo ET(2002) Shiga toxin-associated hemolytic uremic syndrome: absence of recurrence after renal transplantation. *Pediatr Nephrol* 17: 809–814

86 Miller RB, Burke BA, Schmidt WJ, Gillingham KJ, Matas AJ, Mauer M, Kashtan CE (1997) Recurrence of haemolytic-uraemic syndrome in renal transplants: a single-centre report. *Nephrol Dial Transplant* 12: 1425–1430

87 Donne RL, Abbs I, Barany P, Elinder CG, Little M, Conlon P, Goodship TH (2002) Recurrence of hemolytic uremic syndrome after live related renal transplantation associated with subsequent *de novo* disease in the donor. *Am J Kidney Dis* 40: E22

88 Remuzzi G, Ruggenenti P, Codazzi D, Noris M, Caprioli J, Locatelli G, Gridelli B (2002) Combined kidney and liver transplantation for familial haemolytic uraemic syndrome. *Lancet* 359: 1671–1672

89 Remuzzi G, Ruggenenti P, Colledan M, Gridelli B, Bertani A, Bettinaglio P, Bucchioni S, Sonzogni A, Bonanomi E, Sonzogni V et al. (2005) Hemolytic uremic syndrome: a fatal outcome after kidney and liver transplantation performed to correct factor H gene mutation. *Am J Transplant* 5: 1146–1150

Role of complement and Factor H in hemolytic uremic syndrome

Christine Skerka and Mihály Józsi

Department of Infection Biology, Leibniz Institute for Natural Product Research and Infection Biology, Hans Knoell Institute, Beutenbergstr. 11a, 07745 Jena, Germany

Introduction

Defective complement control is a cause for atypical hemolytic uremic syndrome (aHUS). A number of recent studies have identified mutations of several complement regulatory genes in aHUS patients. Mutations have been identified in the genes encoding the fluid-phase complement regulator Factor H, the serine protease Factor I and the surface-bound regulator membrane cofactor protein (MCP/CD46). All three proteins control the activity of the central alternative pathway amplification convertase C3bBb. A total of 65 mutations have been identified in the Factor H gene (Tab. 1), 4 in the Factor I gene [1, 2] and 4 mutations in the gene coding for MCP [3, 4]. In addition, autoantibodies that bind and inactivate the function of the immune regulator Factor H have been reported [5].

The functional analyses of the mutant proteins have defined a protective role of Factor H, Factor I and MCP for the integrity of host endothelial cells. These analyses are also indicative for the disease mechanism of aHUS, as defective alternative pathway complement regulation causes generation of aggressive complement-activation products that relate to endothelial cell damage.

The majority of the identified Factor H gene mutations are clustered in the C terminus of the protein, which includes the cell surface recognition region (reviewed in [6]). This clustering of mutations in the C terminus indicates a central role of the cell surface recognition domain for pathophysiology of aHUS. Mutations of additional complement regulatory genes, whose products control the activity of the central amplification convertase C3bBb, are associated with aHUS. These scenarios indicate that defective local inhibition of complement, occurring most likely upon an inflammatory reaction, results in endothelial cell damage. Here we discuss how mutated, defective or inactivated complement regulatory proteins relate to the disease aHUS and focus on complement Factor H.

Complement and Kidney Disease, edited by Peter F. Zipfel

Table 1 - Mutations in Factor H gene associated with atypical HUS

No.	Mutation no.	CCP	Mutation	Result	Case		Age onset	Factor H 70–130% 235–750 mg/L	C3 (650–1850 mg/L)	Comment	Reference
1	1	1	155 del 4bp	32stop	Single case	he	36 y	222 mg/L	0.4	Stop	[17]
2	2	1	A305G	R78G	101	he	–	–	–		[33]
3	3	2	445 del 25bp	136 stop	G case 9	he	18 mo	45%	430	Stop	[20]
4	4	7	C1271A	Q400K	N case 16	he	birth	–	–		[20]
5	5	8	1494 delA	479 stop	F40	he	adult	569 mg/L	680	Stop	[19]
6	6	11	T1963G	C630W	Case 1	he	–	774 mg/L	876	C	[32]
7	7	11	G2091A	C673Y	J case 12	he	34	41%	453	C	[20]
8	8	12	C2214G	S714 stop	Case 2	he	–	619 mg/L	562	Stop	[32]
9	9	13	2373+A	774 stop	F case 8	he	16 mo	35%	310	Stop	[20]
10	10	14	G2621A	E850K	Case 3	he	–	927 mg/L	2000	Charge	[32]
11	11	14	G2776T	V835L	Single case	he					Neumann unpub.
12	12	15	G2742T	S890I	F45	he	22 y	830 mg/L	620	+ADAMTS 13 mut	[10]
13	13	15	A2751G	H893R	I case 11	he	9 mo	30%	240	Charge	[20]
14	14	15	T2770A	Y899stop	Single case	ho	8 mo	<1%	260	Stop	[21]
15		15	T2770A	Y899stop	A case1,2	ho	7 mo	<1%	50	Stop	[20]
16	15	15	T2816A	C915S	H case 10	he	26	45%	270	C	[20]
17	16	15	C2846T	Q924stop	E case 7	he	6 mo	50%	580	Stop	[20]
18	17	15	G2850T	C926F	Single case	he	30 mo	Low	520	C	[82]
19	18	16	G2923T+T2924C	Q950H+Y951H	118	he	–	–	Low	Charge	[33]
20	19	16	T2924C	Y951H	087	he	–	–	Low		[33]
21	20	16	C2940T	T956M	HUS12	he	13 y	821 mg/L	1350		[70]
22	21	16	G3007T	W978C	Case 4	he	–	703 mg/L	754	C	[32]
23	22	17	A3032G	T987A	Single case		–	–	–	–	Neumann unpub.
24	23	17 +20	A3135T+ C3701T	Y1021F + R1210C	Case 16	he	–	787 mg/L	749	C, charge	[32]

25	24	17	T3200C	C1043R	Case 5	he	–	560 mg/L	748	C, charge	[32]
26	25	18	C3299G	Q1076E	Case 6	he	–	511 mg/L	1820	Charge	[32]
27	26	18/19	C3299G, 3559 delA	Q1076E 1168 stop	Pat. 1	he	–	–	–	Stop	[3]
28	27	19	A3429G	D1119G	Pat. 2	he	–	Normal	Normal	Charge	[3]
29	28	19	T3474G	V1134G	Case 7	he	–	835 mg/L	835		[32]
30	29	19	T3497G	Y1142D	Case 8	he	–	1007 mg/L	425	Charge	[32]
31	30	19	T3542C	W1157R	Case 9	he	–	894 mg/L	508	Charge	[32]
32	31	19	C3562G	C1163W	R087	he	–	–	low	C	[33]
33	32	19	3566 +1 G/A	Splice defect	Case 10	he	–	729 mg/L	826	Stop	[32]
34	33	20	G3587T	E1172 stop	022	he	–	–	Mod low	Stop	[33]
35		20	G3587T	E1172 stop	081	he	–	–	mod. low	Stop	[33]
36		20		E1172 stop	R043	he	–	–		Stop	Heinen unpub.
37	34	20	G3619T	R1182S	Single case					Charge	Neumann unpub.
38	35	20	T3620C	W1183R	Case 11	he	–	619 mg/l	890	Charge	[32]
39	36	20	T3620A	W1183R	Single case	he	2 y	normal	normal	Charge	[83]
40	37	20	G3621T	W1183L	L case 14	he	4 mo	94%	470		[20]
41		20	G3621T	W1183L	HUS2	he	23 y	1160 mg/l	710	Red bi C3b	[70]
42	38	20	C3624G	T1184R	Pat. 4	he	–	normal	–	Charge	[71]
43	39	20	C3638T	L1189F	HUS 37						[84]
44	40	20	T3639G	L1189R	HUS11	he	–	1225 mg/l	1670 mg/l	Charge	[70]
45	41	20	C3645G	S1191W	HUS 30						[84]
46	42	20	C3645T, T3663C	S1191L; V1197A	Pat. 5	he	–	–	–		[3]
47	43	20	G3654A	G1194D	130	he	–	–	mod. low	Charge	[33]
48		20		G1194D	single case	–	–	–	–	Charge	[60]
49	44	20	G3654 + 36	G1194+ 12 aa	Single case	he	12 m0	520 mg/l	800 mg/l	Deform.	[72]
50	45	20	T3663C	V1197A	056	he	–	–	–		[33]
51	46	20	T3663C	V1197A+Null all.	HUS3	he	53y	110 mg/l	590	Red bi C3b	[70]
52	47	20	A3666C	E1198A	152	he	–	–	–	Charge	[33]
53	48	20	T3669C	F1199S	K case 13	he	29 y	37%	382		[20]
54	49	20	C3701T	R1210C	HUS29	he	14 mo	198 mg/l E	710	Red bi C3b	[74]

Table 1 – continued

No.	Mutation no.	CCP	Mutation	Result	Case		Age onset	Factor H 70–130% 235–750 mg/L	C3 (650–1850 mg/L)	Comment	Reference
55		20		R1210C	Single case						[60]
56		20	C3701T	R1210C	Case 12	he	–	580 mg/l	1250		[32]
57		20	C3701T	R1210C	F106	he	we –8 y	711 mg/l	610		[19]
58		20	C3701T	R1210C	R16	he	48 y	731 mg/l	1220		[19]
59		20	C3701T	R1210C	069	he	–	–	mod. low		[33]
60	50	20	C3716G	R1215G	2a IV12	–	–	486 mg/l	750	Charge	[17]
61	51	20	G3717A	R1215Q	F34	he	we –8 y	561 mg/l	690	Stop	[19]
62	52	20	3719 del ACA	1216 del T	Case 13	he	–	–	–		[32]
63	54	20	C3749T	P1226S	Case 14	he	–	1300 mg/l	1020	Red bind	[32]
64	53	20	A3747Tdel 24bp	1228stop	F85 Bedouin	ho	we –8 y	<70 mg/l	100	Stop	[19]
65	55	20	3768 del AGAA	Frameshift + 37 aa	Case 15	he	–	503 mg/l	540	Longer transcript	[32]

he, heterozygous mutation; ho, homozygous mutation; A, alanine; C, cysteine; E, glutamic acid; F, phenylalanine; G, glycine; H, histidine; K, lysine; L, leucine; Q, glutamine; R, arginine; S, serine; T, threonine; V, valine; W, tryptophan; Y, tyrosine; red bi, reduced binding. Numbering of base pairs/amino acids is adapted from Ripoche et al 1988 [30]. Factor H and C3 levels are original data as published by the authors.

HUS: the disease

HUS is characterized by microangiopathic hemolytic anemia, thrombocytopenia and acute renal failure, and was first described by Gasser and co-workers in 1955 [7]. Three forms of HUS can be differentiated: an epidemic, sporadic and familial form of HUS.

The epidemic form of HUS is the most common and usually manifests by infections with toxin-producing *Escherichia coli* of the serotype 0157:H7 or *Shigella dysenteria* serotype 1. The virulence of these microbes is associated with the production of shiga toxin or shiga-like toxins SLT1 and SLT2 that inhibit protein synthesis at the ribosomes, and have a direct toxic effect on renal epithelial cells [8]. This form of HUS is common in children, and is often associated with diarrhea and, therefore, also termed D^{+}HUS or Shiga toxin HUS. Most of the patients show acute renal failure; however, renal function recovers frequently [9].

The second and third form, the atypical form (aHUS) can be sporadic and familial occurring HUS cases, which usually are not associated with diarrhea and are also termed D–HUS or non-shiga toxin HUS. The sporadic form can be triggered by genetic defects, neuraminidase produced by *S. pneumoniae*, or by other factors, such as pregnancy, drugs like cyclosporin A, diseases such as systemic lupus erythematosus (SLE) or infection with the human immune deficiency virus (HIV).

The familial form is a rare disease but carries a significant burden of morbidity and mortality. Patients often suffer from end-stage renal failure and depend on continuous dialysis [10].

Complement and aHUS

Reduced plasma levels of the central complement serum protein C3 were reported already in 1975 in both sporadic and familial forms of aHUS [11]. In addition, the presence of C3 cleavage products have been identified in plasma of aHUS patients. These parameters are indicative either of defective complement control, or of an inherited defect in C3 protein synthesis. The role of defective complement was underlined by reduced C3 levels between relapses [12]. Granular C3 deposits were also observed in glomeruli and arterioles in aHUS patients and enhanced levels of complement cleavage products (C3a and C5a) in serum were indicative of uncontrolled complement activation due to defective control. In 1981 Thompson and Winterborn [13] described a young aHUS patient with very low C3 plasma levels and low levels of complement Factor H. This study identified a defective complement control and Factor H as disease causing factor, and initiated further investigations. Several additional studies showed decreased levels or abnormal mobilities of the Factor H protein in SDS-PAGE, or decreased C3 plasma levels and/or complement functional activities for patients with familial and sporadic HUS [12, 14–16].

By genetic linkage analysis in three large pedigrees of familial aHUS, Warwicker and co-workers [17] identified in 1998 the regulators of complement activation (RCA) gene cluster on chromosome 1 as a hot spot that segregated with aHUS. Consequently, the authors reported mutations in the Factor H gene in patients with aHUS [18]. Several genetic studies reported mutations in the Factor H gene in aHUS patients, and genetic analyses identified mutations in two other genes coding for complement regulatory proteins, i.e., the serine protease Factor I and the membrane-bound regulator MCP. The three gene products Factor H, MCP and Factor I directly control or affect the activity of the alternative complement pathway amplification convertase C3bBb.

aHUS-associated Factor H gene mutations

Up to now 65 mutations have been identified in the Factor H gene that are associated with aHUS (Fig. 1 and Tab. 1). The identified mutations represent either nonsense mutations that introduce a premature stop codon, or missense mutations that result in single amino acid exchanges, deletions or insertions. The vast majority of the mutations are heterozygous, thus the patients have one defective and one intact allele. Three exceptions to this rule exists. One was described for a Bedouin family, which has a premature stop codon in complement control protein (CCP) domain 20, that causes deletion of the four C-terminal amino acids [19]. A truncated Factor H protein was expressed and the potentially functionally defective protein was found

Figure 1
Domain structure of complement Factor H and location of HUS-associated mutations
(A) Factor H is composed of 20 CCP (or short consensus repeats) domains. The individual domains are numbered consecutively. The N-terminal complement regulatory region (represented by CCPs 1–4) is shown in yellow and the C-terminal recognition region in blue. The location of the C3 and heparin/polyanion binding regions are shown above the structure by horizontal lines. The position of the reported Factor H gene mutations in HUS are given below the structure. Mutations that introduce premature stop codons are shown in white letters with a black background. Mutations that cause exchange of single amino acids, or that introduce amino acid deletions are shown with black letters. (B) Folding of a single CCP domain, here the sequence of the most C-terminal domain (CCP 20) is shown. The conserved cysteine residues (C) are shown in blue and the disulfide bonds that are formed between Cys I and Cys III and between Cys II and Cys IV are indicated by the blue line. This arrangement forms three loop regions which are termed loop I, loop II and loop III. (C) Sequence of CCP region 19 and 20. The conserved cysteine residues are shown in white letters on a gray background. Amino acids that are mutated in HUS patients are shown in red.

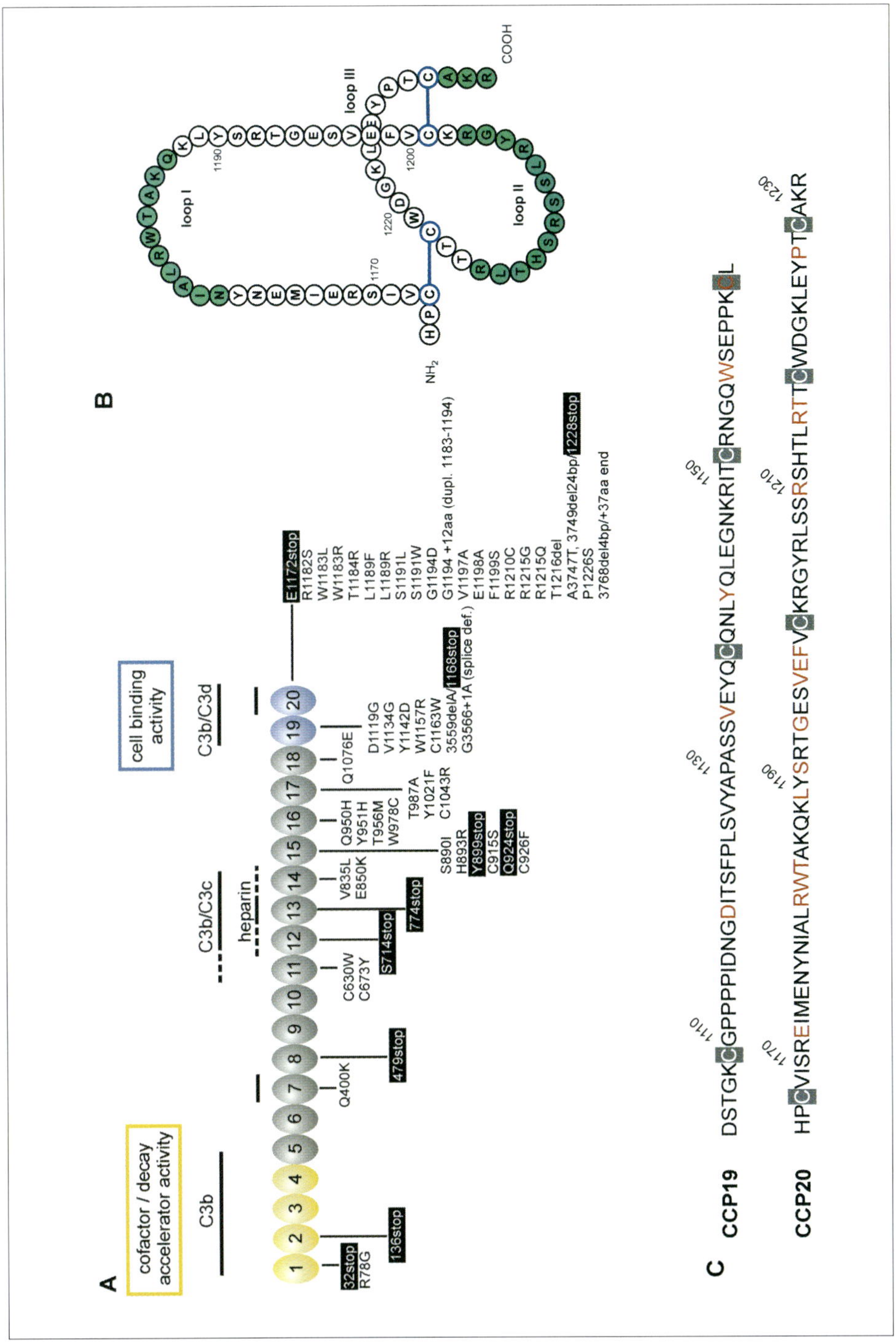
A
cofactor / decay accelerator activity
cell binding activity
C3b
C3b/C3c
heparin
C3b/C3d
1 2 3 4 5 6 7 8 9 10 11 12 13 14 15 16 17 18 19 20
32stop
R78G
136stop
Q400K
479stop
C630W
C673Y
S714stop
774stop
V835L
E850K
S890I
H893R
Y899stop
C915S
Q924stop
C926F
Q950H
Y951H
T956M
W978C
T987A
Y1021F
C1043R
Q1076E
D1119G
V1134G
Y1142D
W1157R
C1163W
3559delA/1168stop
G3566+1A (splice def.)
E1172stop
R1182S
W1183L
W1183R
T1184R
L1189F
L1189R
S1191L
S1191W
G1194D
G1194 +12aa (dupl. 1183-1194)
V1197A
E1198A
F1199S
R1210C
R1215G
R1215Q
T1216del
A3747T, 3749del24bp/1228stop
P1226S
3768del4bp/+37aa end
B
loop I
loop II
loop III
NH2
COOH
C
CCP19 DSTGKCGPPPIDNGDITSFPLSVYAPASSVEYQCQNLYQLEGNKRITCRNGQWSEPPKCL
CCP20 HPCVISREIMENYNIALRWTAKQKLYSRTGESVEFVCKRGYRLSSRSHTLRTTCWDGKLEYPTCAKR

in severely reduced levels in patients plasma (below 70 mg/L). The other homozygous mutations have both a premature stop in CCP 15 of Factor H, and are associated with low Factor H plasma levels [20, 21].

Structure and function of Factor H

Factor H genetics

The human *Factor H* gene (CFH) is located on the long arm of chromosome 1, within the RCA gene cluster in 1q32 [22, 23]. This cluster includes two groups of complement regulatory genes that are separated by a stretch of DNA coding for non-complement genes. The individual RCA genes are probably derived from one ancestral gene, and they have exons coding for a so-called CCP module, also known as short consensus repeat. These modules are mostly encoded by symmetric, phase 1-1 exons, which can be duplicated, inserted, deleted and shuffled. The two groups of RCA genes reflect their evolutionary relationship [24]. The FH gene is found in a group that contains six Factor H family genes: Factor H, the alternatively spliced Factor H gene product FHL-1, and five genes coding for Factor H-related proteins (CFHL1–CFHL5) [25].

The Factor H gene spans 94 kb and contains 23 exons. In general each CCP domain of Factor H is encoded by single exons, with the exception of CCP 2, which represents a split exon and which is encoded by exons III and IV. In addition, the signal peptide and the unique C-terminal region of FHL-1 are encoded by unique exons. The Factor H gene encodes two proteins: Factor H and the Factor H like protein 1 (FHL-1). FHL-1 is derived from an alternatively spliced transcript that covers CCPs 1–7, and an unique C terminus that encodes four amino acids and the 3' untranslated region and that is encoded by exon X [26–28].

Several single nucleotide polymorphisms in the Factor H gene sequence have been observed, both silent ones and those leading to amino acid exchanges [29–33]. Some of these polymorphisms have been associated with aHUS as they are found more often in patients than in healthy controls [20, 33–35].

The Factor H protein

Factor H is a single-chain, 150-kDa plasma glycoprotein. The secreted protein is 1213 amino acids long and is exclusively composed of 20 individual CCP domains (Fig. 1A) [30, 36]. A single CCP domain is about 60 amino acids long, and is stabilized by two disulfide bridges. The four conserved framework cysteine residues form disulfide bridges in a Cys I-Cys III and Cys II-Cys IV manner and allow a globular structure with six β-strands and variable exposed loops [37]. The consecutive

CCP domains are linked by inter domain regions, that are 3–8 amino acids long (residues counted between the conserved Cys IV and Cys I residues). The typical CCP-type folding generates three loop regions (Fig. 1B). The longest loop, i.e., loop I is formed between Cys I and Cys II, loop II represents the amino acid stretch between Cys II and Cys III and loop III is formed between Cys III and Cys IV. In addition to the central framework Cys residues, additional non-framework amino acid residues are conserved in the various CCP domains [27]. These conserved residues are considered relevant for the structure and architecture of an individually folding CCP unit. In contrast, the variable residues, particularly those located in the loop regions, are considered surface exposed and responsible for interaction with the specific ligands. This composition of conserved amino acid residues that form the framework or scaffold of a CCP unit and the variable ligand interacting residues defines the complex architecture of a CCP unit. Such an architecture explains why single amino acid mutations as observed in patients with aHUS [and in membranoproliferative glomerulonephritis (MPGN) patients] can seriously affect both the structure and the function of a CCP composed protein such as the Factor H and the MCP protein.

Factor H functions

Factor H is an essential fluid-phase regulator of the alternative pathway of complement [38–41] and the plasma protein (i) competes with factor B for C3b binding, i.e., prevents the formation of alternative pathway convertases C3bBb (Fig. 2A); (ii) promotes the disassembly of the alternative pathway convertases C3bBb and C3bBbC3b (decay-accelerator activity) (Fig. 2B); and (iii) acts as a cofactor for the inactivation of C3b by the serine protease Factor I (Fig. 2C). This capacity to block the C3b amplification convertase makes Factor H a potent inhibitor of the alternative cascade.

Factor H has two important functional domains that are located at the opposite ends of the protein: The N-terminal recognition region, i.e., CCPs 1–4 include the complement regulatory activities, and the C-terminal recognition domain is contained within CCP domains 19 and 20. Factor H acts as fluid-phase complement regulator and, in addition, with its C-terminal domain, Factor H binds to cell and tissue surfaces, and thus mediates its protective role also on host cell surfaces (Fig. 1) [42–44].

Factor H ligands and ligand binding

Factor H is a multifunctional protein and mediates its regulatory role by interaction with several ligands such as C3b/C3d [45–47], heparin/glycosaminoglycans [48–55],

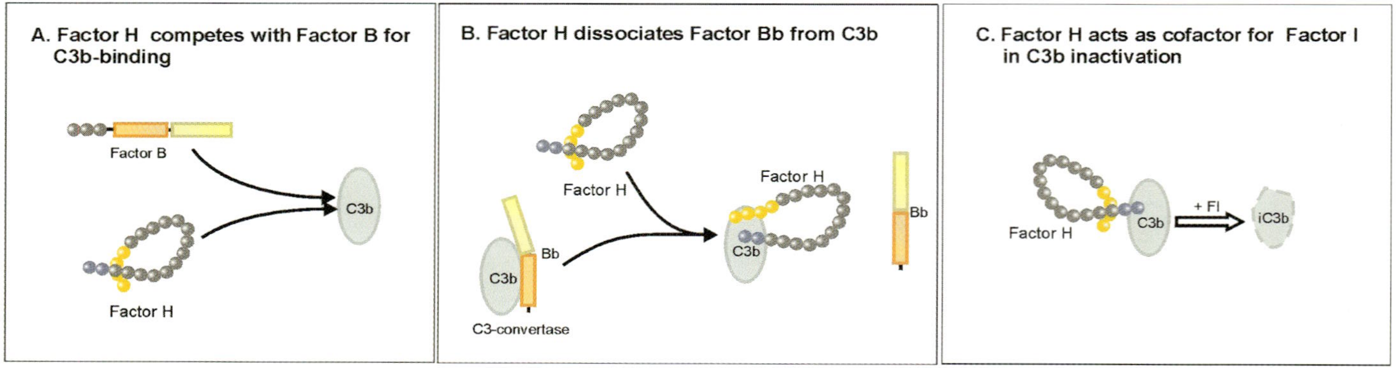

Figure 2

Functions of complement Factor H in control of the amplification convertase C3bBb

Factor H controls the activity and stability of the alternative complement pathway convertase C3bBb. Factor H prevents formation of the C3 convertase by competing with Factor B for C3b binding (A); Factor H facilitates the decay of the C3 convertase by displacing bound Bb (B), and Factor H acts as one of the Factor I cofactors to degrade C3b (C) as shown here for the activity in fluid phase.

adrenomedullin [56], C-reactive protein [57], and multiple bacterial and microbial ligands (reviewed in [58, 59]) (Fig. 1). Apparently, C3b and heparin represent important ligands of Factor H, as the Factor H protein has three binding sites for both. Heparin binding regions are localized to CCP 7, to a region around CCP 13 and to CCP 20. The domains represented by CCP 19 and CCP 20 is of special interest and intensively investigated as it includes several binding regions like that for binding to cell surfaces. Homology modeling [60], mutational analysis and peptide spot analysis identified amino acids relevant for heparin binding and the visualization of a surface-exposed contact region. A first biochemical-based mutational analysis identified residues K1108 in CCP 19 and K1202, R1203, R1206, R1210, K1222, K1230 and R1231 in CCP 20 as central elements for heparin binding (Fig. 1C) [61]. A recombinant Factor H 15-20 fragment with mutated residues (R1203E, R1206E, R1210S, K1230S and R1231A) showed severely reduced binding to heparin, thus confirming that these positively charged amino acids mediated binding to heparin. Ganesh et al. [62] determined the structure of vaccinia virus control protein (VCP) complexed with a heparin decamer fragment, and used this structure for identification of amino acids relevant for Factor H–heparin interaction. Upon superimposition of CCP 20 of Factor H with CCP 4 of VCP, again the basic residues R1182, K1202, R1203, R1206, R1210, R1215, K1230 and R1231 were in contact with heparin. Hybridization experiments with a fluorescently labeled heparin molecule to short peptides representing the complete sequence of CCP 19 and 20 identified four linear heparin binding sites, which include the positively charged residues identified in the former studies: region 1 is located in CCP 19 (aa 1136–1157), region 2 (aa 1178–1187) and region 3 (aa 1202–1215) in CCP 20 and in the C-terminal region of Factor H (aa 1230 and 1231) [63].

The C3 binding sites have been mapped within the Factor H protein to CCPs 1–4, to CCPs 11–14 (C3b and C3c) and to CCPs 19 and 20 (C3b and C3d) [61, 64–66]. The C-terminal binding sites of Factor H for C3b and C3d overlap with the heparin binding site. This has been demonstrated using heparin as an inhibitor for binding of Factor H fragment 15–20 to C3b [61] and by peptide spot analysis, showing overlapping binding sites for all three ligands, heparin, C3b and C3d [63].

Factor H mutations

In general terms, the reported aHUS-associated mutations in the Factor H protein [19–21, 67–72] can be grouped into three categories (Fig. 3): (i), nonsense mutations (Fig. 3: scenario I), (ii) missense mutations (Fig. 3: scenarios II and III); and (iii) deletion, insertion or duplication of one or more nucleotides.

(i) Nonsense mutations: In seven cases (i.e., 11%) nonsense mutations that cause introduction of a premature stop codon have been identified. These mutations

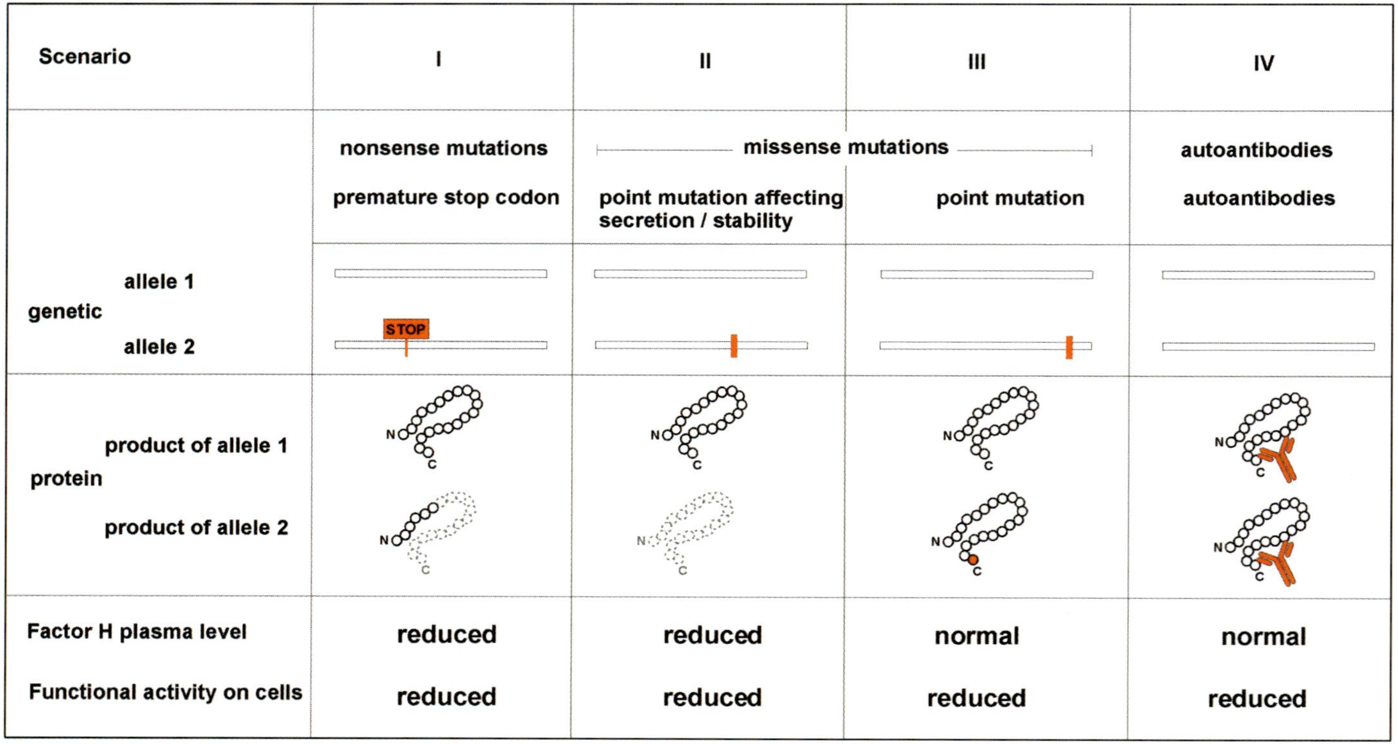
Scenario
I
II
III
IV
nonsense mutations
missense mutations
autoantibodies
premature stop codon
point mutation affecting secretion / stability
point mutation
autoantibodies
genetic
allele 1
allele 2
STOP
protein
product of allele 1
product of allele 2
N
C
Factor H plasma level
reduced
reduced
normal
normal
Functional activity on cells
reduced
reduced
reduced
reduced

have been reported for CCP 15 (three separate mutations) and CCP 20 (CCP 12, three separate mutations) (Fig. 1 and Tab. 1). It is currently unclear whether these truncated proteins are expressed and secreted. Plasma levels of Factor H of these patients are mostly reduced.

(ii) Missense mutations: The vast majority of mutations represent single amino acid exchanges (70%). These mutations cluster in the C terminus of the Factor H protein and are located in the recognition domain of the protein in CCP 19 and 20 (46%). Based on results of the plasma levels and of Factor H gene mutations reported and characterized in MPGN (see Chapter by Zipfel, Smith and Heinen) it is hypothesized that these single amino acid mutations either affect secretion of the defective allelic product (a), or result in secretion of a functionally defective protein (b).

(a) Mutations of both framework and non-framework residues cause a block of protein secretion as has been clearly shown for MPGN type II. In these cases the mutated protein is expressed intracellularly and is retained in the cytoplasm. As the vast majority of aHUS patients are heterozygous for the mutation, it is difficult to assess defective protein secretion. Factor H plasma levels vary in neonates 170–400 µg/mL, in children 225–1600 µg/mL and in adults 235–750 µg/mL [73]. In addition, various laboratories use different methods for the determination of Factor H concentrations in plasma, and therefore it is currently impossible to decide which of the reported mutations is associated with a defect in protein secretion.

(b) The second type of missense mutations produce a secreted protein product that has a single amino acid exchanged. Several of these mutant Factor H proteins can be identified by their altered mobility by SDS-PAGE in combination with Western blot analysis [43, 74]. Based on the unique structure of the CCP domain, amino acid residues can have a role as scaffold elements and maintain the proper architecture of the domain, or if the over-

Figure 3
Different scenarios of Factor H mutations and modifications that are associated with HUS
The vast majority of Factor H gene mutations are heterozygous, the patients have one defective and one intact allele. The mutations have different effects. Scenario I: The majority of nonsense mutations that introduce premature stop codons seem to result in non-expressed alleles (exception to this are patients who have a premature stop codon in domain 20 such as the Bedouin family). Scenario II: Missense mutations can either cause defective protein secretion or result in a secreted but functionally defective protein (scenario III). Additional scenarios, such as the presence of autoantibodies result in normal Factor H protein, normal secretion and normal Factor H plasma levels, but the function of the protein is impaired (scenario IV).

all structure is intact exchanged amino acids may be located in accessible region within a ligand binding site. Recent studies with recombinant Factor H fragments underline the sensitivity of minor amino acid changes on the complement regulatory functions of Factor H and the importance of linker regions [63, 75, 76].

(iii) Deletion, insertion or duplication of one or more amino acids are described for nine aHUS patients (15%). The mutations are spread over the complete Factor H gene and in six cases resulted in a premature stop of translation. The remaining three mutations presumably create Factor H molecules with additional amino acids, which may interfere with normal regulatory functions.

Clustering of mutations in CCP 19 and 20

The mutations identified in the Factor H gene are distributed over the Factor H gene but most of the mutations cluster in the C terminus, particularly in CCP domain 19 and 20 of Factor H, which represent a hot spot. Interestingly nine mutations at five positions have been identified in several unrelated individuals (Tab. 1).

A total of 39 of the reported 65 aHUS-associated mutations in the Factor H gene (i.e., 60%) are located in CCP 19 or CCP 20, and represent either single amino acid mutations, deletions, premature Stop codons or insertions (Tab. 1, Fig. 1A). This clustering of mutations within the two most C-terminal CCP domains of the flexible Factor H protein highlights the role of the C-terminal recognition region for Factor H physiology.

The importance of the C-terminal recognition region for proper Factor H functions on cell surfaces was shown experimentally using mutant Factor H purified from plasma of aHUS patients, such as Factor H E1172Stop [43, 77] or in studies with recombinantly expressed fragments of Factor H that carry the aHUS-associated mutations R1210C, R1215G [43, 74], W1157R, W1183L, V1197A and P1226S [63, 74], or recombinant Factor H proteins deleted of the C-terminal CCPs (Fig. 1C) [42]. These studies demonstrated severely reduced binding of the mutated proteins to cell surfaces, and suggest that reduced or defective complement control on cell surfaces is associated with aHUS.

16 different single amino acid exchanges in CCP 20 of Factor H have been identified in aHUS patients. Thus, it is of interest to understand how these different mutations result into the same functional defect. Considering a role as compositorial scaffold residues, which maintain the overall architecture, or a role as directly interacting residues, exchanges of single amino acids can have different effects. They may either affect the architecture of a single CCP domain, or, if the architecture remains intact, the exchange may affect ligand binding regions and, thus, directly interfere with ligand binding, e.g., C3b/C3d and heparin/glycosaminoglycan. This

scenario was confirmed in a detailed comparison and localization of a number of aHUS-associated mutations [63]. Amino acid residues in CCP 20 of Factor H that are conserved among all members of the Factor H protein family seem responsible for the architecture of the globular structure. Mutations of these amino acids may affect the general architecture, and therefore influence the orientation of a complete loop. Thus, a complete binding region may be affected and become inverted, tilted and inaccessible. However, if the overall structure remains intact, an exchange of an amino acid located within a central surface-exposed position may directly interfere with ligand interaction. Consequently, ligand binding such as C3b/C3d or heparin/glycosaminoglycans is impaired. In addition, as shown for the Factor H protein with the R1210C mutation, the mutant protein changes its characteristics and interacts with other proteins like albumin [74].

The currently available data show that CCP 19 and 20 of Factor H mediate a central role in recognition and bind heparin and C3b/C3d, and represent the major binding domain for cell surfaces [42–44, 61, 63, 74, 77, 78]. The C-terminal heparin binding region is rather sensitive for mutations and modifications, and so far no polymorphic variants have been described for this region. Consequently, it appears that any mutation within this region of Factor H does directly interfere with ligand and/or cell binding. Defective cell binding blocks the protective role of Factor H on the surface of cells [44, 77, 78].

Factor H mutations and C3b/d binding sites

In addition to affecting heparin binding, mutations in Factor H also change the binding characteristics to C3b and C3d. Several binding domains for C3b/d have been described in Factor H. The complement regulatory domain in CCP 1–4 harbors a C3b binding site [75], but also CCP 11–14 and CCP 19/20 can bind C3b/d [61, 64, 65, 79]. Mutations in CCP 19/20 also affect C3b and C3d binding, as demonstrated in binding studies with recombinant mutant proteins [43, 63, 74]. These data are in agreement with overlapping binding sites for heparin and C3b/d determined in the C terminus [61, 63].

Autoantibodies to Factor H in aHUS patients

In addition, autoantibodies against Factor H have been identified in plasma of three patients with recurrent aHUS. These antibodies affected Factor H function and caused decreased Factor H activity. This observation provides evidence for an acquired functional Factor H deficiency, supporting the fact that aHUS occurs in a context of an autoimmune disease (Fig. 3: scenario IV) [5].

Mutations in other complement regulatory proteins: Factor I and MCP (CD46)

Serine protease Factor I

Human complement Factor I is a serine protease that in the presence of appropriate cofactors cleaves the α chain of C3b and inhibits the alternative pathway amplification loop. Thus, recently identified mutations in the gene coding for Factor I in patients suffering from aHUS confirm the role of impaired complement regulation in the pathogenesis of this disease [1, 2].

Membrane cofactor protein

Complement control particularly at the cell surface is mediated by multiple regulators of the fluid-phase and the integral membrane regulators such as MCP (CD46), complement receptor 1 (CR1/CD35) and decay-accelerating factor (DAF/CD55). Recently mutations in the gene coding for the MCP have been identified in four families with aHUS [3, 4, 80].

MCP is a surface-bound membrane regulator, is widely expressed on almost every human cell, and has four CCP domains that mediate the complement regulatory cofactor activity of the protein (reviewed in [81]). The protein also contains a serine/proline rich domain, a transmembrane and an intracellular cytoplasmic domain. So far, eight aHUS cases from four families have been reported with a mutated MCP gene. These mutations include a deletion of two amino acids (D237 and S238 in CCP domain 4; two individuals from family I) and an exchange of S206P in CCP domain 4 (two patients of family II are heterozygous, and two patients of family III are homozygous for this mutation) [3]. In another family, a frame shift results in an exchange of three amino acids (X233, Y234 and Z235) and a premature stop codon that causes the loss of the transmembrane domain [4]. All these mutations caused reduced surface expression, reduced binding of C3b and/or reduced function of MCP at the cell membrane. Thus, the mutant protein causes reduced complement control at the surface of the cells. MCP is expressed in high levels at glomerular cells (MCP-associated mutations are discussed in detail in the chapter by Goodship et al.)

Conclusions

The alternative pathway of complement is spontaneously and continuously activated, and clinical, genetic and molecular data show that defective complement control is associated with aHUS. Particularly on host cells and tissue surfaces, activation of the amplification convertase C3bBb is strongly controlled and continuously inhibited. The activity of the initial convertase and that of the amplification convertase is tightly controlled and regulated by a combination of surface-attached regulators such as Factor H and FHL-1, and by integral membrane proteins such as CR1, MCP and DAF. The importance of the regulation and inhibition of the aggressive complement activation products is demonstrated by the severity of diseases that occur when the activity of these regulators is unbalanced. Under situations of stress, a combined local action of the multiple functionally identical regulators is required to maintain integrity of host cell surfaces. Particularly during stress and high-level complement activation, a reduced activity or a defect of one single regulator may shift the balance from protection to damage. Such a scenario explains why the alternative pathway regulators Factor H and MCP as well as the serine protease Factor I, which in the presence of a cofactor inactivates the newly formed surface-deposited C3b molecule, are all associated with aHUS. In contrast, a complete loss of Factor H functions by homozygous mutations that cause autosomal dominant mutations lead to MPGN type II early in life (see chapter by Zipfel et al).

Although considerable clinical and genetic evidence shows an association of complement regulation with aHUS and MPGN type II, the molecular mechanisms of how these functional disturbances induce cell damage, particularly of the renal endothelial cell, and cause the diseases are not fully understood. In addition, the initial triggers that lead to complement activation are not clearly identified. The following aspects support the view that the damage to the endothelium by complement is a cause for the development of aHUS:

(i) Factor H displays its complement regulatory functions in both the fluid phase and on self cell surfaces (Figs 2 and 4).
(ii) A total of 60% of the reported aHUS mutations in the Factor H gene cluster are located in the C terminus of the protein and impair the recognition function of Factor H. A mutant Factor H protein may show normal complement regulatory activity in fluid phase. However, the impaired cell binding activity may lead to reduced protection at the cell surface, e.g., endothelial cells.
(iii) The remaining 27% of the reported aHUS mutations in the Factor H gene are located in CCPs 1–18. Presumably these mutations interfere directly with proper complement regulatory functions as these mutations, for instance, replace cysteines that are important for the structure of the molecule, or introduce charged amino acid residues into the sequence. Thus, these Factor H mutant proteins might bind to cell surfaces but have impaired complement regulatory functions.
(iv) Reduced Factor H plasma levels by autoantibodies against complement Factor H or a defect in secretion reduce plasma protein levels of Factor H, and result in a reduced protection of endothelial cells.
(v) Mutations in the membrane-bound complement regulator MCP result in reduced protection of the cell surfaces.

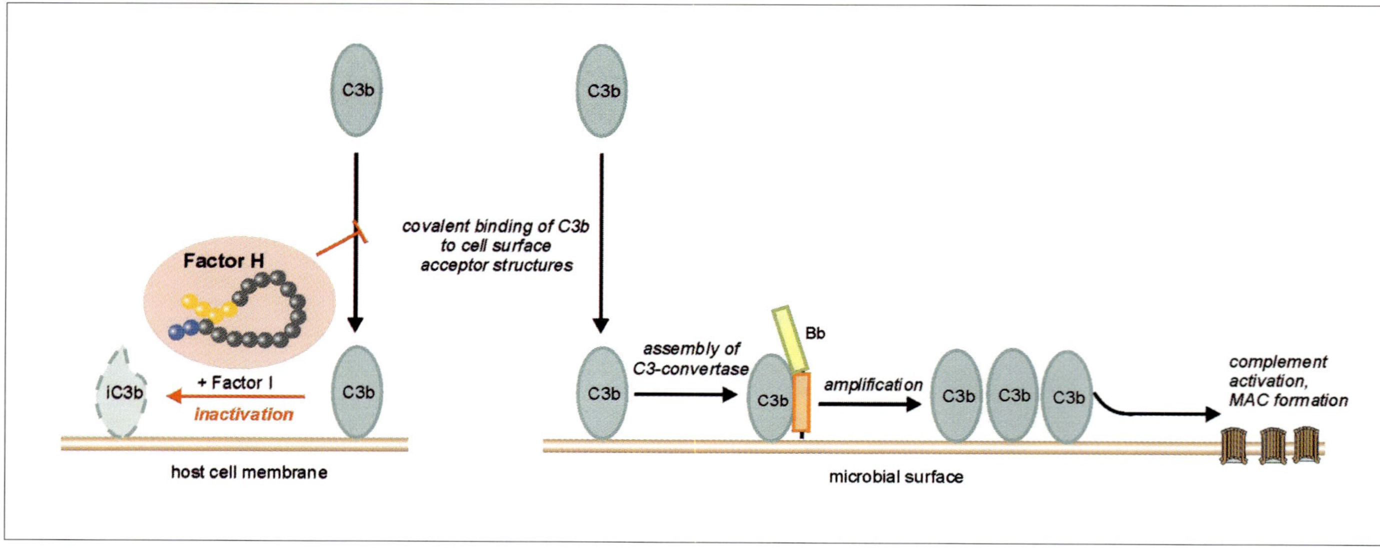

Figure 4
Alternative complement pathway activation on various surfaces
The alternative pathway of complement is activated spontaneously and continuously. If the pathway is not inhibited, activation proceeds and a powerful amplification convertase is formed, e.g., on an activator surface such as microbe (right panel). In contrast, on host cells that represent a non-activator surface, the cascade is blocked and C3b is inactivated by the serine protease Factor I. Factor H acts as a major cofactor for this protease and also acts on the surface of host cells.

(vi) Mutations in serine protease Factor I lead to complement activation on cell surfaces, as Factor H and membrane-bound complement regulators require Factor I for degradation of the C3 convertase C3bBb.
(vii) Verotoxin-producing *E. coli* which might induce the typical form of HUS act directly on endothelial cells and cause damages of the cell lining.
(viii) A complement-mediated damage of the endothelial cell lining can induce the formation of thrombi that leads to microangiopathies.

As the alternative complement pathway is continuously activated at a very low level, endothelial cells always need to be protected by efficient regulators like Factor H or MCP. Plasma Factor H is recruited to the cell surfaces and fulfils two major functions, binding to nonactivator surfaces together with a proper complement regulatory function. These properties are even more important in cases of infections when microorganisms enhance local complement activation, which produces damaging activation products that need to be properly controlled and inactivated on the surface of bystander cells. In this situation the endothelial cell needs maximum protection and actually exploits a concert of several complement regulators.

Apparently having a half normal level of intact Factor H or MCP is not sufficient to protect endothelial kidney cells in patients with heterozygous mutations. In addition, several individual exist who carry the mutations but are healthy and show no signs of the disease. Thus, most likely an additional factor like stress, microbial infection, drugs or injuries may initiate an inflammatory response that causes endothelial cell damage and may eventuate to aHUS.

References

1 Fremeaux-Bacchi V, Dragon-Durey MA, Blouin J, Vigneau C, Kuypers D, Boudailliez B, Loirat C, Rondeau E, Fridman WH (2004) Complement factor I: a susceptibility gene for atypical haemolytic uraemic syndrome. *J Med Genet* 41: e84

2 Kavanagh D, Kemp EJ, Mayland E, Winney RJ, Duffield JS, Warwick G, Richards A, Ward R, Goodship JA, Goodship TH (2005) Mutations in complement Factor I predispose to development of atypical hemolytic uremic syndrome. *J Am Soc Nephrol* 16: 2150–2155

3 Richards A, Kemp EJ, Liszewski MK, Goodship JA, Lampe AK, Decorte R, Muslumanoglu MH, Kavukcu S, Filler G, Pirson Y et al (2003) Mutations in human complement regulator, membrane cofactor protein (CD46), predispose to development of familial hemolytic uremic syndrome. *Proc Natl Acad Sci USA* 100: 12966–12971

4 Noris M, Brioschi S, Caprioli J, Todeschini M, Bresin E, Porrati F, Gamba S, Remuzzi G; International Registry of Recurrent and Familial HUS/TTP (2003) Familial haemolytic uraemic syndrome and an MCP mutation. *Lancet* 362: 1542–1547

5 Dragon-Durey MA, Loirat C, Cloarec S, Macher MA, Blouin J, Nivet H, Weiss L, Fridman WH, Fremeaux-Bacchi V (2005) Anti-Factor H autoantibodies associated with atypical hemolytic uremic syndrome. *J Am Soc Nephrol* 16: 555–563

6 Zipfel PF, Neumann HP, Józsi M (2003) Genetic screening in haemolytic uraemic syndrome. *Curr Opin Nephrol Hypertens* 12: 653–657

7 Gasser C, Gautier E, Steck A, Siebenmann RE, Dechslin R (1955) Hämolytisch-urämische syndrome bilaterale Nierenrindennekrosen bei akuten erworbenen hämolytischen Anämien. *Schweiz Med Wschr* 85: 905–909

8 Moake JL (2001) Thrombotic Microangiopathies. *N Engl J Med* 347: 589–600

9 Siegler R, Oakes R (2005) Hemolytic uremic syndrome; pathogenesis, treatment, and outcome. Curr Opin Pediatr 17: 200–204

10 Noris M, Remuzzi G (2005) Hemolytic uremic syndrome. *J Am Soc Nephrol* 16:1035–1050

11 Kaplan BS, Chesney RW, Drummond KN (1975) Hemolytic uremic syndrome in families. *N Engl J Med* 292: 1090–1093

12 Noris M, Ruggenenti P, Perna A, Orisio S, Caprioli J, Skerka C, Vasile B, Zipfel PF, Remuzzi G (1999) Hypocomplementemia discloses genetic predisposition to hemolytic uremic syndrome and thrombotic thrombocytopenic purpura: role of factor H abnormalities. Italian Registry of Familial and Recurrent Hemolytic Uremic Syndrome/ Thrombotic Thrombocytopenic Purpura. *J Am Soc Nephrol* 10: 281–293

13 Thompson RA, Winterborn MH (1981) Hypocomplementemia due to a genetic deficiency of b1H globulin. *Clin Exp Immunol* 46: 110–119

14 Pichette V, Querin, S, Schürich W, Brun G, Lehner-Netsch G, Delaghe J-M (1994) Familial hemolytic-uremic syndrome and homozygous factor H deficiency. *Am J Kidney Dis* 26: 936–941

15 Rougier N, Kazatchkine MD, Rougier JP, Fremeaux-Bacchi V, Blouin J, Deschenes G, Soto B, Baudouin V, Pautard B, Proesmans W et al (1998) Human complement factor H deficiency associated with hemolytic uremic syndrome. *J Am Soc Nephrol* 9: 2318–2326

16 Ohali M, Shalev H, Schlesinger M, Katz Y, Kachko L, Carmi R, Sofer S, Landau D (1998) Hypocomplementemic autosomal recessive hemolytic uremic syndrome with decreased factor H. *Pediatr Nephrol* 12: 619–624

17 Warwicker P, Goodship TH, Donne RL, Pirson Y, Nicholls A, Ward RM, Turnpenny P, Goodship JA (1998) Genetic studies into inherited and sporadic hemolytic uremic syndrome. *Kidney Int* 53: 836–844

18 Warwicker P, Donne RL, Goodship JA, Goodship TH, Howie AJ, Kumararatne DS, Thompson RA, Taylor CM (1999) Familial relapsing haemolytic uraemic syndrome and complement factor H deficiency. *Nephrol Dial Transplant* 14: 1229–1233

19 Caprioli J, Bettinaglio P, Zipfel PF, Amadei B, Daina E, Gamba S, Skerka C, Marziliano N, Remuzzi G, Noris M; Italian Registry of Familial and Recurrent HUS/TTP (2001) The molecular basis of familial hemolytic uremic syndrome: mutation analysis of factor H gene reveals a hot spot in short consensus repeat 20. *J Am Soc Nephrol* 12: 297–307

20 Dragon-Durey MA, Fremeaux-Bacchi V, Loirat C, Blouin J, Niaudet P, Deschenes G, Coppo P, Herman Fridman W, Weiss L (2004) Heterozygous and homozygous factor H deficiencies associated with hemolytic uremic syndrome or membranoproliferative glomerulonephritis: report and genetic analysis of 16 cases. *J Am Soc Nephrol* 15: 787–795
21 Licht C, Weyersberg A, Heinen S, Stapenhorst L, Devenge J, Beck B, Waldherr R, Kirschfink M, Zipfel PF, Hoppe B (2005) Successful plasma therapy for atypical hemolytic uremic syndrome caused by factor H deficiency owing to a novel mutation in the complement cofactor protein domain 15. *Am J Kidney Dis* 45: 415–421
22 Rodríguez de Córdoba S, Díaz-Guillén MA, Heine-Suñer D (1999) An integrated map of the human regulator of complement activation (RCA) gene cluster on 1q32. *Mol Immunol* 36: 803–808
23 Díaz-Guillén MA, Rodríguez de Córdoba S, Heine-Suñer D (1999) A radiation hybrid map of complement factor H and factor H-related genes. *Immunogenetics* 49: 549–552
24 Krushkal J, Bat O, Gigli I (2000) Evolutionary relationships among proteins encoded by the regulator of complement activation gene cluster. *Mol Biol Evol* 17: 1718–1730
25 Józsi M, Richter H, Loschmann J, Skerka C, Buck F, Beisiegel U, Erdei A, Zipfel PF (2005) FHR-4A: a new factor H-related protein is encoded by the human FHR-4 gene. *Eur J Hum Genet* 13: 321–329
26 Schwaeble W, Zwirner J, Schulz TF, Linke RP, Dierich MP, Weiss EH (1987) Human complement factor H: expression of an additional truncated gene product of 43 kDa in human liver. *Eur J Immunol* 17: 1485–1489
27 Zipfel PF, Skerka C (1994) Complement factor H and related proteins: an expanding family of complement-regulatory proteins? *Immunol Today* 15: 121–126
28 Friese MA, Hellwage J, Jokiranta TS, Meri S, Peter HH, Eibel H, Zipfel PF (1999) FHL-1/reconectin and factor H: two human complement regulators which are encoded by the same gene are differently expressed and regulated. *Mol Immunol* 36: 809–818
29 Rodríguez de Córdoba S, Rubinstein P (1984) Genetic polymorphism of human factor H (β1H). *J Immunol* 132: 1906–1908
30 Ripoche J, Day AJ, Harris TJ, Sim RB (1988) The complete amino acid sequence of human complement factor H. *Biochem J* 249: 593–602
31 Warwicker P, Goodship THJ, Goodship JA (1997) Three new polymorphisms in the human complement factor H gene and promoter region. *Immunogenetics* 46: 437–438
32 Neumann HP, Salzmann M, Bohnert-Iwan B, Mannuelian T, Skerka C, Lenk D, Bender BU, Cybulla M, Riegler P, Konigsrainer A et al (2003) Haemolytic uraemic syndrome and mutations of the factor H gene: a registry-based study of German speaking countries. *J Med Genet* 40: 676–681
33 Caprioli J, Castelletti F, Bucchioni S, Bettinaglio P, Bresin E, Pianetti G, Gamba S, Brioschi S, Daina E, Remuzzi G, Noris M (2003) International Registry of Recurrent and Familial HUS/TTP. Complement factor H mutations and gene polymorphisms in haemolytic uraemic syndrome: the C-257T, the A2089G and the G2881T polymorphisms are strongly associated with the disease. *Hum Mol Genet* 12: 3385–3395

34 Fremeaux-Bacchi V, Kemp EJ, Goodship JA, Dragon-Durey MA, Strain L, Loirat C, Deng HW, Goodship TH (2005) The development of atypical HUS is influenced by susceptibility factors in factor H and membrane cofactor protein- evidence from two independent cohorts. *J Med Genet* 2005 Mar 22; [Epub ahead of print]

35 Esparza-Gordillo J, Goicoechea de Jorge E, Buil A, Carreras Berges L, Lopez-Trascasa M, Sanchez-Corral P, Rodriguez de Cordoba S (2005) Predisposition to atypical hemolytic uremic syndrome involves the concurrence of different susceptibility alleles in the regulators of complement activation gene cluster in 1q32. *Hum Mol Genet* 14: 703–712

36 Kristensen T, Tack BF (1986) Murine protein H is comprised of 20 repeating units, 61 amino acids in length. *Proc Natl Acad Sci USA* 83: 3963–3967

37 Norman DG, Barlow PN, Baron M, Day AJ, Sim RB, Campbell D (1991) Three dimensional structure of a complement control protein module in solution. *J Mol Biol* 219: 717–725

38 Nilsson UR, Müller-Eberhard HJ (1965) Isolation of β1F-globulin from human serum and its characterization as the fifth component of complement. *J Exp Med* 122: 277–298

39 Whaley K, Ruddy S (1976) Modulation of the alternative complement pathway by beta 1H globulin. *J Exp Med* 144: 1147–1163

40 Weiler JM, Daha MR, Austen KF, Fearon DT (1976) Control of the amplification convertase of complement by the plasma protein beta1H. *Proc Natl Acad Sci USA* 73: 3268–3272

41 Pangburn MK, Schreiber RD, Müller-Eberhard HJ (1977) Human complement C3b inactivator: isolation, characterization, and demonstration of an absolute requirement for the serum protein beta1H for cleavage of C3b and C4b in solution. *J Exp Med* 146: 257–270

42 Pangburn MK (2002) Cutting edge: localization of the host recognition functions of complement factor H at the carboxyl-terminal: implications for hemolytic uremic syndrome. *J Immunol* 169: 4702–4706

43 Manuelian T, Hellwage J, Meri S, Caprioli J, Noris M, Heinen S, Józsi M, Neumann HP, Remuzzi G, Zipfel PF (2003) Mutations in factor H reduce binding affinity to C3b and heparin and surface attachment to endothelial cells in hemolytic uremic syndrome. *J Clin Invest* 111: 1181–1190

44 Józsi M, Manuelian T, Heinen S, Oppermann M, Zipfel PF (2004) Attachment of the soluble complement regulator factor H to cell and tissue surfaces: relevance for pathology. *Histol Histopathol* 19: 251–258

45 Conrad DH, Carlo JR, Ruddy S (1978) Interaction of beta1H globulin with cell-bound C3b: quantitative analysis of binding and influence of alternative pathway components on binding. *J Exp Med* 147: 1792–1805

46 Jokiranta TS, Zipfel PF, Hakulinen J, Kuhn S, Pangburn MK, Tamerius JD, Meri S (1996) Analysis of the recognition mechanism of the alternative pathway of complement

by monoclonal anti-factor H antibodies: evidence for multiple interactions between H and surface bound C3b. *FEBS Lett* 393: 297–302

47 Pangburn MK, Pangburn KL, Koistinen V, Meri S, Sharma AK (2000) Molecular mechanisms of target recognition in an innate immune system: interactions among factor H, C3b, and target in the alternative pathway of human complement. *J Immunol* 164: 4742–4751

48 Fearon DT (1978) Regulation by membrane sialic acid of beta1H-dependent decay-dissociation of amplification C3 convertase of the alternative complement pathway. *Proc Natl Acad Sci USA* 75: 1971–1975

49 Kazatchkine MD, Fearon DT, Austen KF (1979) Human alternative complement pathway: membrane-associated sialic acid regulates the competition between B and beta1 H for cell-bound C3b. *J Immunol* 122: 75–81

50 Meri S, Pangburn MK (1990) Discrimination between activators and nonactivators of the alternative pathway of complement: Regulation via a sialic acid/polyanion binding site on factor H. *Proc Natl Acad Sci USA* 87: 3982–3986

51 Pangburn MK, Atkinson MA, Meri S (1991) Localization of the heparin-binding site on complement factor H. *J Biol Chem* 266: 16847–16853

52 Meri S, Pangburn MK (1994) Regulation of alternative pathway complement activation by glycosaminoglycans: specificity of the polyanion binding site on factor H. *Biochem Biophys Res Commun* 198: 52–59

53 Blackmore TK, Sadlon, TA, Ward, HM, Lublin DM, Gordon, DL (1996) Identification of a heparin binding domain in the seventh short consensus repeat of complement factor H. *J Immunol* 157: 5422–5427

54 Blackmore TK, Hellwage J, Sadlon TA, Higgs N, Zipfel PF, Ward HM, Gordon DL (1998) Identification of the second heparin-binding domain in human complement factor H. *J Immunol* 160: 3342–3348

55 Ram S, Sharma AK, Simpson SD, Gulati S, McQuillen DP, Pangburn MK, Rice PA (1998) A novel sialic acid binding site on factor H mediates serum resistance of sialylated Neisseria gonorrhoeae. *J Exp Med* 187: 743–752

56 Pio R, Martinez A, Unsworth EJ, Kowalak JA, Bengoechea JA, Zipfel PF, Elsasser TH, Cuttitta F (2001) Complement factor H is a serum-binding protein for adrenomedullin, and the resulting complex modulates the bioactivities of both partners. *J Biol Chem* 276: 12292–12300

57 Jarva H, Jokiranta TS, Hellwage J, Zipfel PF, Meri S (1999) Regulation of complement activation by C-reactive protein: targeting the complement inhibitory activity of factor H by an interaction with short consensus repeat domains 7 and 8-11. *J Immunol* 163: 3957–3962

58 Zipfel PF, Hellwage J, Friese MA, Hegasy G, Jokiranta ST, Meri S (1999) Factor H and disease: a complement regulator affects vital body functions. *Mol Immunol* 36: 241–248

59 Lindahl G, Sjobring U, Johnsson E (2001) Human complement regulators: a major target for pathogenic microorganisms. *Curr Opin Immunol* 12: 576–582

60 Perkins SJ, Goodship TH (2002) Molecular modelling of the C-terminal domains of fac-

tor H of human complement: a correlation between haemolytic uraemic syndrome and a predicted heparin binding site. *J Mol Biol* 316: 217–224

61 Hellwage J, Jokiranta TS, Friese MA, Wolk TU, Kampen E, Zipfel PF, Meri S (2002). Complement C3b/C3d and cell surface polyanions are recognized by overlapping binding sites on the most carboxyl-terminal domain of complement factor H. *J Immunol* 169: 6935–6944

62 Ganesh VK, Smith SA, Kotwal GJ, Murthy KH. (2004) Structure of vaccinia complement protein in complex with heparin and potential implications for complement regulation. *Proc Natl Acad Sci USA* 101: 8924–8929

63 Jozsi M, Heinen S, Hartmann A, Ostrowicz CD, Hälbich S, Richter H, Kunert A, Licht C, Saunders RE, Perkins SJ et al (2005) Factor H and atypical hemolytic uremic syndrome: mutations in the C-terminus cause structural changes and defective recognition functions. *JASN; in press*

64 Sharma AK, Pangburn MK (1996) Identification of three physically and functionally distinct binding sites for C3b in human complement factor H by deletion mutagenesis. *Proc Natl Acad Sci USA* 93: 10996–11001

65 Jokiranta TS, Hellwage J, Koistinen V, Zipfel PF, Meri S (2000) Each of the three binding sites on complement factor H interacts with a distinct site on C3b. *J Biol Chem* 275: 27657–27662

66 Oppermann M, Manuelian T, Józsi M, Brandt E, Hellwage J, Jokiranta TS, Meri S, Skerka C, Götze O, Zipfel PF (2005) The C-terminus of complement regulator Factor H is central for target recognition: Evidence for a compact conformation of the native protein. *Submitted*

67 Ault BH, Schmidt BZ, Fowler NL, Kashtan CE, Ahmed AE, Vogt BA, Colten HR (1997) Human factor H deficiency. *J Biol Chem* 272: 25168–25175

68 Ying L, Katz Y, Schlesinger M, Carmi R, Shalev H, Haider N, Beck G, Sheffield VC, Landau D (1999) Complement factor H gene mutation associated with autosomal recessive atypical hemolytic uremic syndrome. *Am J Hum Genet* 65: 1538–1546

69 Buddles MR, Donne RL, Richards A, Goodship J, Goodship TH (2000) Complement factor H gene mutation associated with autosomal recessive atypical hemolytic uremic syndrome. *Am J Hum Genet* 66: 1721–1722

70 Perez-Caballero D, Gonzalez-Rubio C, Gallardo ME, Vera M, Lopez-Trascasa M, Rodriguez de Cordoba S, Sanchez-Corral P (2001) Clustering of missense mutations in the C-terminal region of factor H in atypical hemolytic uremic syndrome. *Am J Hum Genet* 68: 478–484

71 Richards A, Buddles MR, Donne RL, Kaplan BS, Kirk E, Venning MC, Tielemans CL, Goodship JA, Goodship TH (2001) Factor H mutations in hemolytic uremic syndrome cluster in exons 18-20, a domain important for host cell recognition. *Am J Hum Genet* 68: 485–490

72 Filler G, Radhakrishnan S, Strain L, Hill A, Knoll G, Goodship TH (2004) Challenges in the management of infantile factor H associated hemolytic uremic syndrome. *Pediatr Nephrol* 19: 908–911

73 de Paula PF, Barbosa JE, Junior PR, Ferriani VP, Latorre MR, Nudelman V, Isaac L (2003) Ontogeny of complement regulatory proteins - concentrations of factor H, factor I, C4b-binding protein, properdin and vitronectin in healthy children of different ages and in adults. *Scand J Immunol* 58: 572–577

74 Sanchez-Corral P, Perez-Caballero D, Huarte O, Simckes AM, Goicoechea E, Lopez-Trascasa M, Rodriguez de Cordoba S (2002) Structural and functional characterization of factor H mutations associated with atypical hemolytic uremic syndrome. *Ann J Hum Genet* 71: 1285–1295

75 Kühn S, Skerka C, Zipfel PF (1995) Mapping of the complement regulatory domains in the human factor H-like protein 1 and in factor H1. *J Immunol* 155: 5663–5670

76 Lehtinen MJ, Meri S, Jokiranta TS (2004) Interdomain contact regions and angles between adjacent short consensus repeat domains. *J Mol Biol* 344: 1385–1396

77 Heinen S, Józsi M, Noris M, Remuzzi G, Zipfel PF (2006) An HUS associated mutant Factor H protein that lacks domain 20 shows reduced cofactor activity on the surface of endothelial cells. *Mol Immunol* 43: 165–166 [abstract]

78 Sanchez-Corral P, Gonzalez-Rubio C, Rodriguez de Cordoba S, Lopez-Trascasa M (2004) Functional analysis in serum from atypical hemolytic uremic syndrome patients reveals impaired protection of host cells associated with mutations in factor H. *Mol Immunol* 41: 81–84

79 Jokiranta TS, Westin J, Nilsson UR, Nilsson B, Hellwage J, Lofas S, Gordon DL, Ekdahl KN, Meri S (2001) Complement C3b interactions studied with surface plasmon resonance technique. *Int Immunopharmacol* 1: 495–506

80 Goodship TH, Liszewski MK, Kemp EJ, Richards A, Atkinson JP (2004) Mutations in CD46, a complement regulatory protein, predispose to atypical HUS. *Trends Mol Med* 10: 226–231

81 Liszewski MK, Farries TC, Lublin DM, Rooney IA, Atkinson JP (1996) Control of the complement system. *Adv Immunol* 61: 201–283

82 Cheong HI, Lee BS, Kang HG, Hahn H, Suh KS, Ha IS, Choi Y (2004) Attempted treatment of factor H deficiency by liver transplantation. *Pediatr Nephrol* 19: 454–458

83 Remuzzi G, Ruggenenti P, Codazzi D, Noris M, Caprioli J, Locatelli G, Gridelli B (2002) Combined kidney and liver transplantation for familial haemolytic uraemic syndrome. *Lancet* 359: 1671–1672

84 Rodriguez de Cordoba S, Esparza-Gordillo J, Goicoechea de Jorge E, Lopez-Trascasa M, Sanchez-Corral P (2004) The human complement factor H: functional roles, genetic variations and disease associations. *Mol Immunol* 41: 355–367

Genetic testing in atypical HUS and the role of membrane cofactor protein (MCP; CD46) and Factor I

Timothy H.J. Goodship[1], *Véronique Frémeaux-Bacchi*[2], *John P. Atkinson*[3]

[1]Institute of Human Genetics, University of Newcastle upon Tyne, Newcastle upon Tyne, NE1 3BZ, UK; [2]Hôpital Européen Georges Pompidou, 20 rue Leblanc, 75908 Paris Cedex 15, France; [3]Division of Rheumatology, Washington University School of Medicine, St Louis, MO 63110, USA

Introduction

There has been a dramatic increase in our understanding of the molecular mechanisms responsible for those diseases that are characterized by a thrombotic microangiopathy, including atypical hemolytic uremic syndrome (aHUS). In aHUS, the finding of mutations in the plasma complement regulatory protein Factor H [1] has now been extended by the description of mutations in the transmembrane complement regulator, membrane cofactor protein (MCP, CD46) [2] and the serine protease Factor I [3]. These three proteins work in concert to regulate the complement cascade at the step of C3 activation, and thus implicate this system in the pathophysiology of HUS. Understanding the molecular mechanisms of these diseases now affords the opportunity to develop logical and more targeted treatment. In this chapter we focus on recent observations pertaining to MCP and Factor I, and discuss the role of genetic testing in aHUS.

Atypical HUS

HUS consists of the triad of a microangiopathic hemolytic anemia, thrombocytopenia and acute renal failure. Most commonly this syndrome is associated with a preceding diarrheal illness usually due to infection with *Escherichia coli* O157. Less frequently, there is no preceding diarrheal illness, and hence this form of the disease is known as "atypical HUS". This can either be sporadic or, when more than one member of a family is affected, familial. In both sporadic and familial HUS the disease tends to be recurrent. The clinical features of the syndrome are due to the development of platelet-rich microthrombi in small vessels. This particularly affects the glomeruli of the kidney causing acute renal failure. Recovery of renal function is uncommon in aHUS, and many patients require long-term renal replacement therapy. While dialysis is often successful, recurrence of the disease in renal alllografts is frequently seen.

Complement and Kidney Disease, edited by Peter F. Zipfel

Membrane cofactor protein

MCP was identified in 1985 during an examination of C3b binding proteins expressed by human peripheral blood mononuclear cells (PBMCs) [4]. MCP is expressed on almost every human cell except erythrocytes [5–9]. It was initially called gp45-70 because of its broad two-band electrophoretic mobility pattern. Subsequently, it was discovered to be a cofactor for the liver-synthesized serine protease Factor I and renamed [10]. Together with the serine protease Factor I, it degrades C3b and C4b bound to the same cell surface on which MCP is expressed [10–15]. This proteolytic inactivation event is known as cofactor activity.

MCP consists of four alternatively spliced isoforms that co-exist on most cells (Fig. 1) [16–18]. On gel electrophoresis, MCP appears as two broad, heterogeneous bands [4, 10]. The extracellular domain is composed of four complement control protein modules (CCPs). Following these repeats is an O-glycosylated region rich in serines, threonines and prolines, the "STP region", which is alternatively spliced. There are either 14 or 29 amino acids in the STP region, and this depends on the splicing in or out of STP exon B. This results in a characteristic structure of MCP on SDS-PAGE with two species of 51–58 kDa ("lower band") and 59–68 kDa ("upper band"). The presence of STP-B produces the higher and its absence the lower molecular mass isoform [17]. The STP region is followed by a juxtamembranous region of 12 amino acids of unknown function followed by a hydrophobic transmembrane domain and a charged cytoplasmic anchor. The cytoplasmic tail of MCP is also alternatively spliced (the 16 amino acid CYT-1 and the 23 amino acid CYT-2), and mediates signaling events [16, 19]. There are four commonly expressed isoforms: MCP-BC1, MCP-BC2, MCP-C1 and MCP-C2, where B and C refer to the STP region and 1 or 2 the cytoplasmic tail. The *MCP* gene is located within the regulators of complement activation (RCA) cluster on the long arm of chromosome 1 at 3.2 and consists of 14 exons spanning ~43 kb.

Function of MCP

The human complement system promotes the inflammatory response and modifies membranes through opsonization and lytic activity. It is part of both the innate and adaptive immune systems. Three pathways of complement activation lead to generation of C3b, which deposits on the target (Fig. 2). That nearly half of the proteins in the system are regulators emphasizes the importance of controlling the system. In most cases, foreign surfaces usually lack membrane-bound regulators and are unable to bind soluble regulators, so they are attacked by complement. Host cells are protected from complement activation by regulatory proteins that inactivate C3b deposited on their surface [20, 21]. As noted above, they accomplish this through a cofactor protein which must bind C3b or C4b before Factor I can cleave.

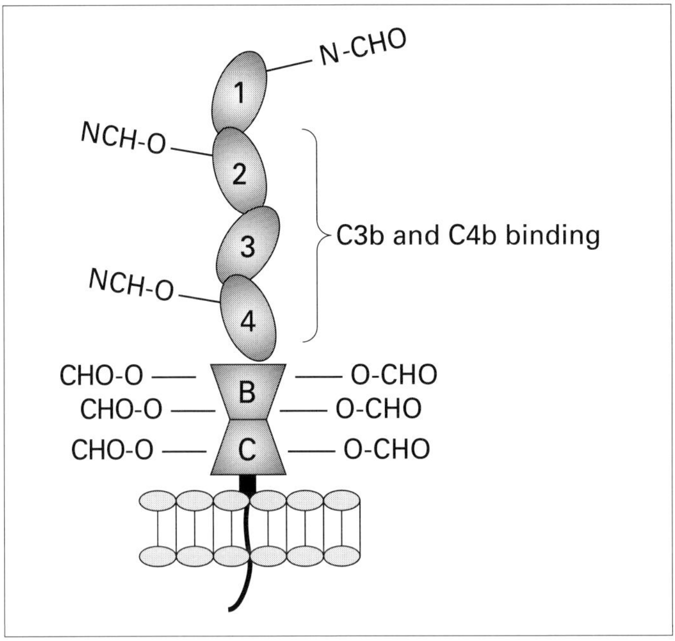

Figure 1
Schematic diagram of the structure of MCP
The amino portion (shown in blue) consists of four complement control protein repeats, each about 60 amino acids in length, that contains three N-linked sugars. This is followed by a domain bearing O-linked sugars termed B and C. The exon encoding the B region is alternatively spliced. A transmembrane hydrophobic domain and one of two possible alternatively spliced cytoplasmic tails completes the protein at its C terminus (reproduced from [54] with permission from Elsevier) .

There are five principle regulators of complement activation at the steps of C3 and C5 cleavage; the enzymes involved in these reactions are known as convertases. C4 binding protein (C4BP) and Factor H are plasma proteins that regulate the classical and alternative pathway, respectively, by both decay-accelerating activity and cofactor activity. Decay-accelerating factor (DAF, CD55), MCP (CD46) and complement receptor 1 (CR1, CD35) are membrane bound and regulate both the classical and alternative pathways. DAF possess only decay-accelerating and MCP only cofactor activity (Fig. 3), while the much less widely expressed CR1 possesses both. DAF and MCP are synergistic in their mode of action [22]. Studies with deletion mutants and monoclonal antibodies (mAbs) have shown that ligand binding and cofactor activity are contained in CCPs 2, 3 and 4 of MCP [23, 24]. Besides its

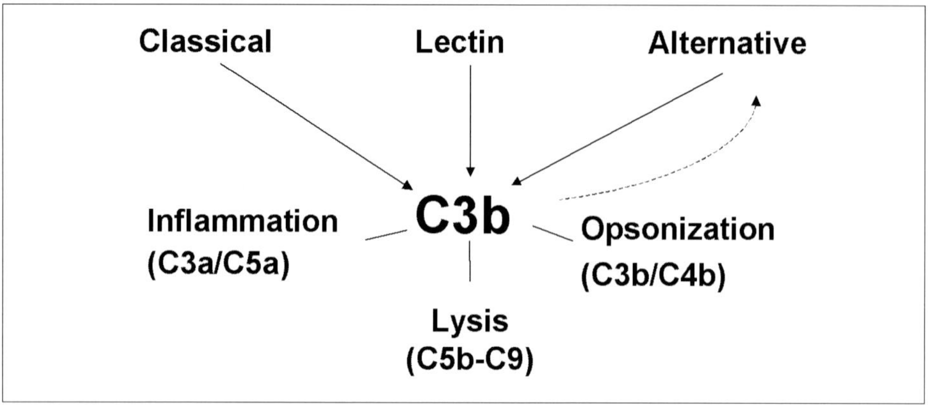

Figure 2
The three pathways of complement activation
Deposition of clusters of C3b on a target is the primary goal. As shown by the broken line, the alternative pathway also serves as a feedback loop to amplify C3b deposition (reproduced from [54] with permission from Elsevier) .

cofactor and ligand binding activity, MCP is a receptor for the measles virus, *Streptococcus pyogenes*, pathogenic Neisseria, herpes virus 6 and some of the adenoviruses [25, 26].

MCP and HUS

In an early study we established linkage in three HUS families to the RCA cluster on chromosome at 1q32 [27]. At that time the first candidate gene we examined was Factor H because of previous case reports associating *Factor H* deficiency with HUS [28–30]. However, in only one of the three families did we find a *Factor H* mutation. Because *MCP* lies within the RCA cluster on chromosome 1, we looked at it as a candidate gene in the other two families. In one family we detected in affected individuals a heterozygous 6 base pair deletion (GACAGT) in the exon coding for CCP 4, resulting in the loss of two amino acids (ΔD237/S238). In this family three male siblings were affected at the ages of 27, 31 and 35 years [31]. C3 levels at presentation were normal, and there was no recovery of renal function. Subsequently, all three received a cadaveric renal transplant with no recurrence of HUS in the allograft. Flow cytometric analysis of PBMCs from an affected individual showed reduced expression of MCP, and C3b binding studies of PBMC cell lysates revealed ~50% of expected C3b binding activity. Mutant *MCP* constructs were evaluated after transfection into CHO cells. Western blot, flow cytometry, cell surface label-

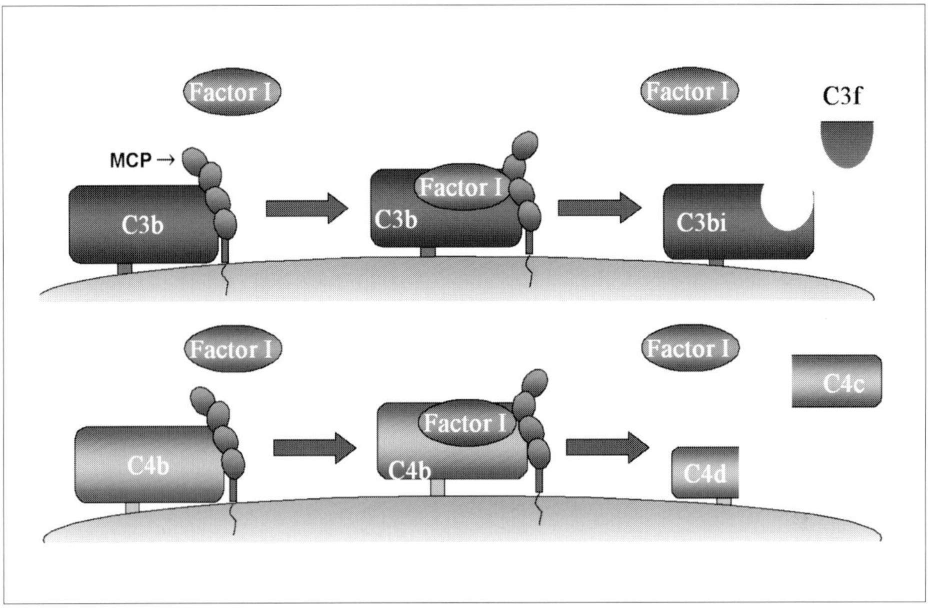

Figure 3
Complement regulatory function of MCP
MCP binds to C3b or C4b attached to the same cell on which it is expressed. The plasma serine protease factor I can then cleave these proteins, thereby preventing further complement activation. This regulatory process is known as cofactor activity. Factor I returns to the plasma, while MCP remains on the membrane ready to bind to more deposited C3b. Factor H possesses cofactor activity C3b (reproduced from [54] with permission from Elsevier).

ing (biotinylation) and pulse chase analysis all showed that the mutant protein was retained intracellularly. Noris et al. [32] have also described an HUS patient with a heterozygous mutation in the exon coding for CCP 4 that results in a premature stop. This was associated with reduced expression of MCP on PBMCs.

The finding of the *MCP* deletion mutation (ΔD237/S238) led us to examine a panel of other HUS families in whom we had not found a *Factor H* mutation. In two families, we found a transition (T822C) resulting in a serine to proline change, S206P, in the exon coding for CCP 4. In one family two male siblings were affected at the ages of 8 and 15 years. In both renal function recovered. The two affected individuals in this family were heterozygous for the S206P substitution. In the other family, one male and one female sibling were affected. Both again made a complete recovery. The affected individuals in this family were homozygous for the substitution, their parents being first degree relatives. In the affected individuals from both families, flow cytometry showed normal MCP expression levels in PBMCs. However, C3b binding was ~50% reduced in affected heterozygotes and nearly undetectable in affected homozygotes.

The functional consequences of the S206P substitution were studied in CHO cells expressing this mutant. All the studies undertaken showed that the mutant protein's ability to interact with C3b was severely impaired (~7% of wild type). C3b cofactor activity was similarly impaired. In addition, cells transfected with this mutation and then challenged with a complement activation process, demonstrated less inhibition of C3b deposition. However, C4b binding and cofactor activity were unaffected. This activity profile strongly suggests a major defect in alternative pathway regulation that predisposes an individual to aHUS.

We have previously generated MCP mutants to identify areas in the CCP domains that are important for C3b/C4b binding and cofactor activity [24]. A 205–212 block peptide deletion exhibited a partial decrease in C3b binding, abrogation of cofactor activity and loss of reactivity to the function blocking mAb GB24. Alterations of individual amino acids (changed to alanines) also altered ligand binding and cofactor activity. S206A, however, had no loss of C3b/C4b binding but ~75% decrease in C3b cofactor activity. The specific change in C3b binding and cofactor activity observed with S206P demonstrates how disease-associated mutations can enhance our understanding of the functional sites of molecules such as MCP.

Thus, these mutations in *MCP*, particularly S206P, suggest that a failure to inactivate or regulate C3b mediates disease development. In particular, both Factor H and MCP bind C3b and possess cofactor activity for C3b [16]. Moreover, the only function of Factor I is to serve as a protease in this reaction. Therefore, we hypothesize that Factor H, Factor I and MCP protect against C3 activation on vascular tissue. MCP provides protection due to its expression on endothelial cells. As the glomerular capillary bed is a fenestrated endothelium, the exposed basement membrane supplies a surface rich in polyanions for Factor H binding. Once bound to such a substrate, Factor H may then function like MCP to protect the tissue.
Although not yet published, data from the Spanish, French and Italian cohorts of aHUS patients confirm that 10–20% of patients have MCP mutations which either lead to reduced expression or impaired function.

Factor I

Besides abnormalities of soluble and membrane-bound regulators of complement activation, mutations in *Factor I* have recently been described in aHUS. Factor I is a soluble regulatory serine protease of the complement system and cleaves three peptide bonds in the α-chain of C3b and two bonds in the α-chain of C4b thereby inactivating these proteins. The protein is a heterodimer of about 88 000 Da, which consists of a non-catalytic heavy chain of 50,000 Da, which is linked by a disulfide bond to a catalytic light chain of 38,000 Da. The protein is synthesized as a single chain precursor of 565 amino acids, predominantly in the liver. Four basic amino acids are then excised from the precursor prior to secretion of the heterodimer. Like many of the complement proteins, Factor I has a modular structure (Fig. 4). The heavy chain contains two low-density lipoprotein receptor (LDLr) domains, a CD5 domain, and a module found only in Factor I and complement proteins C6 and C7. The Factor I gene is located on chromosome 4q25 and spans 63 kb [33]. It comprises 13 exons, and there is a strong correlation between the exonic organization of the gene and the modular structure of the protein. The light chain of Factor I, which is the serine proteinase (catalytic) region of the molecule, is encoded in 5 exons. The genomic organization of the enzymic part of Factor I is similar to that of trypsin. The gene structure is unusual though, in that the first exon is small, 86 bp, and it is separated from the rest of the gene by a large first intron of 36 kb. Homozygous Factor I deficiency has been described previously in approximately 30 kindreds. Because of a secondary C3 deficiency, these patients usually present at an early age with pyogenic infections [34, 35]. Fremeaux-Bacchi et al. [3] have recently described three heterozygous mutations in *Factor I* in aHUS. They screened the coding sequence of *Factor I* in six familial and 19 sporadic cases of aHUS. In two cases, heterozygous nonsense mutations led to a premature stop codon (G456X and W528X). Both of these had approximately half normal antigenic Factor I levels. In the third that had normal Factor I levels there was a heterozygous missense mutation in exon 13 leading to the substitution of an aspartic acid residue by valine (D506V). D506 is a highly conserved residue in the catalytic site of Factor I and its substitution is likely to be functionally significant.

Implications of the association between MCP, Factor I and HUS

The finding of mutations in two further CCPs in aHUS provides further evidence that this is a disease of complement dysregulation (Tab. 1). It is likely that other complement genes will be implicated in those patients and families in whom a *Factor H*, *MCP* or *Factor I* mutation has not yet been found. Other candidate genes include *DAF*, *CR1*, *CD59*, *C3* and *Factor B*.

Do mutations in *MCP* and *Factor I* directly cause aHUS? We favor the hypothesis that the mutations predispose to, rather than directly cause, a thrombotic microangiopathy. In this situation endothelial activation secondary to injury is maintained by excessive complement activation [36]. Prothrombotic effects of complement activation on endothelial cells include:

(i) The formation of C3b clusters that serve as ligands for complement receptors on phagocytic cells (PMNs and monocytes).
(ii) Local liberation of C3a and C5a that bind to C3a and C5a receptors on endothelial cells. This attracts inflammatory cells and activates endothelial cells

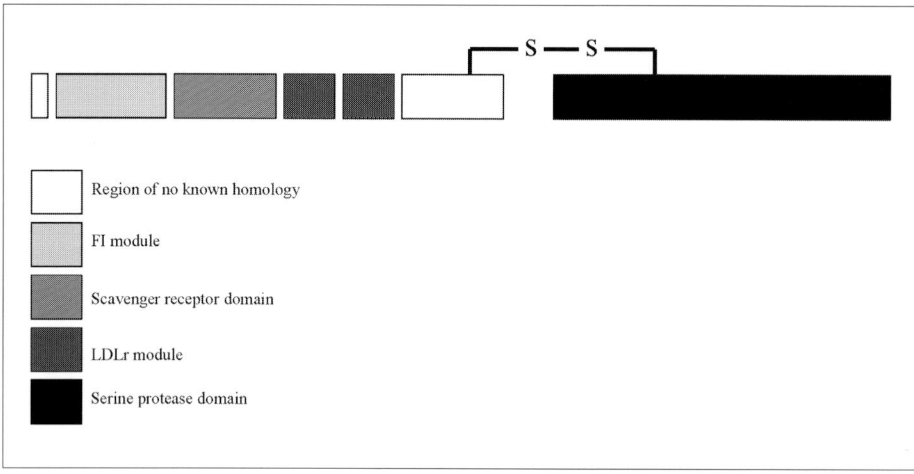

Figure 4
Domain structure of Factor I
(Adapted from [21] with permission from Elsevier).

including the upregulation of procoagulant membrane proteins such as adhesin molecules.

(iii) Formation of the membrane attack complex on the endothelial cell, leading to either cell lysis or sublytic membrane perturbation.

A characteristic feature of Factor H-, MCP- and Factor I-associated HUS is variable penetrance and inheritance. The penetrance of the disease phenotype is approximately 50%. One possible explanation for this is that genetic variation in complement regulatory genes may modify disease penetrance in patients with *Factor H*, *MCP* and *Factor I* mutations. For instance, Caprioli et al. [37] have reported an association between *Factor H* alleles and HUS. This may also explain why HUS has not previously been described in other families with Factor I deficiency.

The MCP and Factor I findings have clinical consequences. Many patients with aHUS require renal replacement therapy. Although peritoneal dialysis and hemodialysis are usually successful in aHUS patients, renal transplantation is complicated by recurrence of the disease in the allograft in approximately 50% of patients [38]. This figure is higher in patients known to have a Factor H abnormality (Tab. 2). Our observation that three patients from the same family with an MCP mutation have been successfully transplanted without recurrence suggests that determination of the genetic defect underlying HUS may facilitate selection of therapy. Based on the finding of mutations in three complement components, we sug-

Table 1 - Abnormalities of complement regulators in aHUS

Regulator	Size (kDa)	Site of synthesis	Function	Heterozygous mutations in aHUS (%)
Factor H (CFH)	150,000	Liver	Binds C3b, CA and DAA for convertases containing C3	20–30
MCP (CD46)	55,000/65,000 (two size variants – arise via alternative splicing)	Most cells (transmembrane protein)	Binds C3b and C4b has CA for both of these ligands	10–20
Factor I (IF)	88,000 (two chains)	Liver	Serine protease – requires a cofactor protein to cleave C3b or C4b	10–20

CA, cofactor activity; DAA, decay accelerating activity

gest that there is sufficient evidence to undertake a clinical trial in some HUS patients of complement inhibitors that block the activation of C3 [39, 40].

Clinical evaluation of the patient with aHUS

The diagnosis of HUS should be suspected in any patient presenting with the aforementioned triad of a microangiopathic hemolytic anemia, thrombocytopenia and acute renal failure. The same triad is also a feature of thrombotic thromboctyopenic purpura (TTP), a syndrome in which neurological and systemic manifestations such as fever are also seen at presentations [41]. Distinction of the two is based on the predominant presenting feature; renal involvement in HUS versus neurological involvement in TTP. The HUS phenotype is associated with abnormalities of complement regulators, whereas TTP is associated with abnormalities of the von-Willebrand Factor cleaving protease ADAMTS13. However, there is overlap between the two. In HUS the complement regulatory abnormality is usually secondary to a primary gene change, while autoimmunity has only been described in three cases with Factor H antibodies [42]. In contrast, the abnormality in ADAMTS13 is usually due to autoantibodies in adults [43] with primary gene changes being described less frequently [44]. The triad of presenting features can also be seen as a secondary manifestation of systemic lupus erythematosus, systemic sclerosis, anti-phospholipid

Table 2 - Reported outcome of transplantation in patients known to have either Factor H *mutations and/or Factor H deficiency*

Reference	n	Outcome	Recurrence
Caprioli et al. [37]	5	All five failed to HUS recurrence within first year	5/5
Olie [47]	1	Two transplants both failed due to recurrence of HUS	2/2
Donne et al. [48] (and unpublished)	2	Two transplants failed due to recurrent HUS	2/2
Fortin [49]	1	Two transplants both failed due to thrombotic microangiopathy/acute allograft glomerulopathy	2/2
Neumann et al. [38]	6	Eight out of ten grafts lost in first year of transplant	8/10
Warwicker et al. [50]	1	No recurrence up to 11 days. Kidney lost due to cortical rupture	0/1
Warwicker et al. [27] (and unpublished)	3	One graft with recurrence on day 24 No recurrence at 7 and 8 years	1/3
Roodhooft et al. [51]	1	HUS recurrence day 27	1/1
Dragon-Durey et al. [52]	5	No recurrence at 2, 4 and 10 years Recurrence at 25 days and 18 months	2/5
Caprioli et al. [53]	3	Recurrence in all three	3/3
Newcastle cohort (unpublished)	3	Recurrence in all three	3/3
Total			29/37 (78%)

antibody syndrome, malignant hypertension, cobalamin C disease and HIV infection. Appropriate investigations should be undertaken to exclude these. An evaluation of complement activation should be undertaken, including measurement in serum of C3, C4, APH50, CH50, Factor I and Factor H. However, it has been shown that *Factor H* and *MCP* mutations can be present without evidence by these types of tests of complement activation. Consideration should therefore be given to undertaking mutation screening of the genes for *Factor H*, *MCP* and *Factor I*.

Molecular evaluation of Factor H, MCP and Factor I

Table 3 summarizes the methods that can be used to detect mutations within or near the genes for Factor H, MCP and Factor I. The two techniques most frequently used for mutation scanning are denaturing high-performance liquid chromatography (DHPLC) and single-strand conformation polymorphism (SSCP). DHPLC is based on heteroduplex analysis where mutant and wild-type sequences present in a PCR

Table 3 - Methods for mutation detection

Method	Advantages	Disadvantages
Southern blot, hybridize to cDNA probe	Only way to determine major deletions and rearrangements	Laborious, expensive Needs several µg DNA
Sequencing	Detects all changes Mutations fully characterized	Expensive
Heteroduplex gel mobility	Very simple Cheap	Sequences <200 bp only Limited sensitivity Does not reveal position of change
DHPLC	Quick, high throughput Quantitative	Expensive equipment Does not reveal position of change
SSCP	Simple, cheap	Sequences < 200 bp only Does not reveal position of change
DGGE	High sensitivity	Choice of primers is critical Expensive primers Does not reveal position of change
Chemical mismatch cleavage	High sensitivity Shows position of change	Toxic chemicals Experimentally difficult
PTT	High sensitivity for chain terminating mutations Shows position of change	Chain terminating mutations only Expensive difficult technique Usually needs RNA
Quantitative PCR	Detects heterozygous deletions	Expensive
Microarrays	Quick High throughput Might detect and define all changes	Expensive Limited range of genes

DHPLC, denaturing high performance liquid chromatography; SSCP, single-strand conformation polymorphism; DGGE, denaturing gradient gel electrophoresis; PTT, protein truncation test. Reproduced with permission from 3rd Edition of Human Molecular Genetics *by T. Strachan and A.P. Read, Garland Science, London.*

product are heated and allowed to re-anneal slowly. This results in the formation of four species of DNA; two heteroduplexes and two homoduplexes. These have altered mobility compared to the wild-type sequence and this is seen as a time shift or variant pattern on DHPLC. In SSCP, the PCR amplicon is denatured, snap-cooled and electrophoresed in a non-denaturing polyacrylamide gel. The single-stranded

DNA assumes a 3D conformation that is dependent on the primary sequence and secondary structure, and this results in altered migration patterns when compared to the wild-type pattern. The disadvantages of these scanning methods are that neither can detect homozygous changes and the sensitivity is not greater than 95%. In addition, all variant patterns have to be sequenced to fully characterize the nature of the change. Sequencing PCR amplicons directly without prior mutation scanning can be costly but the sensitivity of detection approaches 100%. With the wide availability of automated fluorescence sequencers and robotic liquid handling, direct sequencing is increasingly becoming the method of choice for mutation detection. If the exon-specific primers have a 5' N13 (N13 forward GTAGCGCGACGGCCAGT, N13 reverse CAGGGCGCAGCGATGAC) (modified M13) tag then all sequencing reactions can be carried out using a common forward and a common reverse primer. Genomic DNA is sequenced with each exon being amplified with approximately 20 bp of flanking intron. Care has to be taken in designing primers to amplify the exons coding for the C-terminal exons of *Factor H* because of the close homology shared with the Factor H-related proteins [45]. Genomic DNA is extracted from peripheral blood by standard methods and then amplified by PCR using exon-specific primers and annealing temperatures as reported previously [1–3]. PCR products are purified using magnetic microparticles (AMPure, Agencourt Biosciences) to remove unincorporated dNTPs, primers and salts. Sequencing reactions are carried out by dye terminator cycle sequencing using either exon-specific primers or N13 primers, purified using magnetic microparticles (CleanSeq, Agencourt Biosciences) and then electrophoresed on a fluorescence 16 capillary sequencer (Beckman CEQ 8000). The resulting electropherograms are checked for homozygous and heterozygous base changes using an automated sequence analysis package (Mutation Surveyor™, http://www.softgenetics.com). This program has a trace difference component in which the test electopherogram is subtracted from the normal electropherogram and base changes are seen as peaks rising above the background. One disadvantage of both sequencing and mutation scanning, however, is that large deletions and duplications or other large scale rearrangements will not be detected and, therefore, some method of dosage analysis is required. Multiplex ligation-dependent probe amplification (MLPA) is a relatively new method for detecting deletions and duplications in a simple two-stage procedure and gives dosage information for up to 45 exons in one test [46]. Conventional dosage analysis can also be carried out using a fluorescence multiplex PCR but a limited number of exons can be tested in this way.

Summary

Mutations in *Factor H*, *MCP* and *Factor I* have established that aHUS is a disease of complement dysregulation. These findings give hope that complement inhibition may be of therapeutic benefit in this condition. In the near future, we expect mutations to be found in other complement genes. This work has major implications for how we think about syndromes involving essentially any type of tissue injury where the complement cascade is activated (autoantibodies, immune complexes, ischemia/reperfusion injury, sepsis and many others). For a given degree of injury, individuals with defects in Factor H, MCP and Factor I, even if heterozygous, will likely be at increased risk for more severe tissue damage because they cannot appropriately regulate C3 activation and amplification.

Acknowledgements
T.H.J.G. is supported by the National Kidney Research Fund, the Northern Counties Kidney Research Fund, the Peel Medical Research Trust and the Robin Davies Trust. J.P.A. is supported by National Institutes of Health Grant R01 AI37618.

References

1 Richards A, Buddles MR, Donne RL, Kaplan BS, Kirk E, Venning MC, Tielemans CL, Goodship JA, Goodship THJ (2001) Factor H mutations in hemolytic uremic syndrome cluster in exons 18-20, a domain important for host cell recognition. *Am J Hum Genet* 68: 485–490
2 Richards A, Kemp EJ, Liszewski MK, Goodship JA, Lampe AK, Decorte R, Muslumanoglu MH, Kavukcu S, Filler G, Pirson Y et al (2003) Mutations in human complement regulator, membrane cofactor protein (CD46), predispose to development of familial hemolytic uremic syndrome. *Proc Natl Acad Sci USA* 100: 12966–12971
3 Fremeaux-Bacchi V, Dragon-Durey M, Blouin J, Vigneau C, Kuypers D, Boudailliez B, Loirat C, Rondeau E, Fridman W (2004) Complement Factor I: a susceptibility gene for atypical hemolytic-uremic syndrome. *J Med Genet* 41: e84
4 Cole JL, Housley GA Jr, Dykman TR, MacDermott RP, Atkinson JP (1985) Identification of an additional class of C3-binding membrane proteins of human peripheral blood leukocytes and cell lines. *Proc Natl Acad Sci USA* 82: 859–863
5 Endoh M, Yamashina M, Ohi H, Funahashi K, Ikuno T, Yasugi T, Atkinson JP, Okada H (1993) Immunohistochemical demonstration of membrane cofactor protein (MCP) of complement in normal and diseased kidney tissues. *Clin Exp Immunol* 94: 182–188
6 Ichida S, Yuzawa Y, Okada H, Yoshioka K, Matsuo S (1994) Localization of the complement regulatory proteins in the normal human kidney. *Kidney Int* 46: 89–96
7 McNearney T, Ballard L, Seya T, Atkinson JP (1989) Membrane cofactor protein of complement is present on human fibroblast, epithelial, and endothelial cells. *J Clin Invest* 84: 538–545
8 Nakanishi I, Moutabarrik A, Hara T, Hatanaka M, Hayashi T, Syouji T, Okada N, Kitamura E, Tsubakihara Y, Matsumoto M et al (1994) Identification and characteri-

zation of membrane cofactor protein (CD46) in the human kidneys. *Eur J Immunol* 24: 1529–1535

9 Johnstone RW, Loveland BE, McKenzie IF (1993) Identification and quantification of complement regulator CD46 on normal human tissues. *Immunology* 79: 341–347

10 Seya T, Turner JR, Atkinson JP (1986) Purification and characterization of a membrane protein (gp45-70) that is a cofactor for cleavage of C3b and C4b. *J Exp Med* 163: 837–855

11 Barilla-LaBarca ML, Liszewski MK, Lambris JD, Hourcade D, Atkinson JP (2002) Role of membrane cofactor protein (CD46) in regulation of C4b and C3b deposited on cells. *J Immunol* 168: 6298–6304

12 Devaux P, Christiansen D, Fontaine M, Gerlier D (1999) Control of C3b and C5b deposition by CD46 (membrane cofactor protein) after alternative but not classical complement activation. *Eur J Immunol* 29: 815–822

13 Kojima A, Iwata K, Seya T, Matsumoto M, Ariga H, Atkinson JP, Nagasawa S (1993) Membrane cofactor protein (CD46) protects cells predominantly from alternative complement pathway-mediated C3-fragment deposition and cytolysis. *J Immunol* 151: 1519–1527

14 Oglesby TJ, Allen CJ, Liszewski MK, White DJ, Atkinson JP (1992) Membrane cofactor protein (CD46) protects cells from complement-mediated attack by an intrinsic mechanism. *J Exp Med* 175: 1547–1551

15 Seya T and Atkinson JP (1989) Functional properties of membrane cofactor protein of complement. *Biochem J* 264: 581–588

16 Liszewski MK, Farries TC, Lublin DM, Rooney IA, Atkinson JP (1996) Control of the complement system. *Adv Immunol* 61: 201–283

17 Post TW, Liszewski MK, Adams EM, Tedja I, Miller EA, Atkinson JP (1991) Membrane cofactor protein of the complement system: alternative splicing of serine/threonine/proline-rich exons and cytoplasmic tails produces multiple isoforms that correlate with protein phenotype. *J Exp Med* 174: 93–102

18 Russell SM, Sparrow RL, McKenzie IF, Purcell DF (1992) Tissue-specific and allelic expression of the complement regulator CD46 is controlled by alternative splicing. *Eur J Immunol* 22: 1513–1518

19 Riley-Vargus RC and Atkinson JP (2003) Expression of membrane cofactor protein (MCP; CD46) on spermatozoa: Just a complement inhibitor? *Modern Aspects of Immunobiology* 3: 75–78

20 Parker CJ (eds) (1992) *Membrane Defenses Against Attack by Complement and Perforins*. Springer, Berlin

21 Morgan BP, Harris CL (eds) (1999) *Complement Regulatory Proteins*. Academic Press, London

22 Brodbeck WG, Mold C, Atkinson JP, Medof ME (2000) Cooperation between decay-accelerating factor and membrane cofactor protein in protecting cells from autologous complement attack. *J Immunol* 165: 3999–4006

23 Adams EM, Brown MC, Nunge M, Krych M, Atkinson JP (1991) Contribution of the repeating domains of membrane cofactor protein (CD46) of the complement system to ligand binding and cofactor activity. *J Immunol* 147: 3005–3011

24 Liszewski MK, Leung M, Cui W, Subramanian VB, Parkinson J, Barlow PN, Manchester M, Atkinson JP (2000) Dissecting sites important for complement regulatory activity in membrane cofactor protein (MCP; CD46). *J Biol Chem* 275: 37692–37701

25 Segerman A, Atkinson JP, Marttila M, Dennerquist V, Wadell G, Arnberg N (2003) Adenovirus type 11 uses CD46 as a cellular receptor. *J Virol* 77: 9183–9191

26 Gaggar A, Shayakhmetov DM, Lieber A (2003) CD46 is a cellular receptor for group B adenoviruses. *Nat Med* 9: 1408–1412

27 Warwicker P, Goodship THJ, Donne RL, Pirson Y, Nicholls A, Ward RM, Goodship JA (1998) Genetic studies into inherited and sporadic haemolytic uraemic syndrome. *Kidney Int* 53: 836–844

28 Thompson RA and Winterborn MH (1981) Hypocomplementaemia due to a genetic deficiency of beta-1H globulin. *Clin Exp Immunol* 46: 110–119

29 Ohali M, Shalev H, Schlesinger M, Katz Y, Kachko L, Carmi R, Sofer S, Landau D (1998) Hypocomplemententemic autosomal recessive hemolytic uremic syndrome with decreased factor H – Original article. *Pediatr Nephrol* 12: 619–624

30 Pichette V, Querin S, Schurch W, Brun G, Lehnernetsch G, Delage JM (1994) Familial hemolytic-uremic syndrome and homozygous factor-H deficiency. *Am J Kidney Dis* 24: 936–941

31 Pirson Y, Lefebvre C, Arnout C, Van Ypersele de Strihou C (1987) Hemolytic uremic syndrome in three adult siblings: a family study and evolution. *Clin Nephrol* 28: 250–255

32 Noris M, Brioschi S, Caprioli J, Todeschini M, Bresin E, Porrati F, Gamba S, Remuzzi G (2003) Familial haemolytic uraemic syndrome and an MCP mutation. *Lancet* 362: 1542–1547

33 Vyse TJ, Bates GP, Walport MJ, Morley BJ (1994) The organization of the human complement factor I gene (IF): a member of the serine protease gene family. *Genomics* 24: 90–98

34 Vyse TJ, Spath PJ, Davies KA, Morley BJ, Philippe P, Athanassiou P, Giles CM, Walport MJ (1994) Hereditary complement factor I deficiency. *QJM* 87: 385–401

35 Vyse TJ, Morley BJ, Bartok I, Theodoridis EL, Davies KA, Webster AD, Walport MJ (1996) The molecular basis of hereditary complement factor I deficiency. *J Clin Invest* 97: 925–933

36 Ballermann BJ (1998) Endothelial cell activation. *Kidney Int* 53: 1810–1826

37 Caprioli J, Castelletti F, Bucchioni S, Bettinaglio P, Bresin E, Pianetti G, Gamba S, Brioschi S, Daina E, Remuzzi G et al (2003) Complement factor H mutations and gene polymorphisms in haemolytic uraemic syndrome: the C-257T, the A2089G and the G2881T polymorphisms are strongly associated with the disease. *Hum Mol Genet* 12: 3385–3395

38 Neumann HP, Salzmann M, Bohnert-Iwan B, Mannuelian T, Skerka C, Lenk D, Bender BU, Cybulla M, Riegler P, Konigsrainer A et al. (2003) Haemolytic uraemic syndrome and mutations of the factor H gene: a registry-based study of German speaking countries. *J Med Genet* 40: 676–681

39 Kirschfink M (2001) Targeting complement in therapy. *Immunol Rev* 180: 177–189

40 Lambris JD, Holers VM (eds) (2002) *Therapeutic Interventions in the Complement System*. Humana Press, Towata, New Jersey

41 Moake JL (2002) Thrombotic microangiopathies. *N Engl J Med* 347: 589–600

42 Dragon-Durey MA, Loirat C, Cloarec S, Macher MA, Blouin J, Nivet H, Weiss L, Fridman WH, Fremeaux-Bacchi V (2004) Anti-Factor H autoantibodies associated with atypical hemolytic uremic syndrome. *J Am Soc Nephrol* 16: 555–563

43 Tsai H-M, Lian ECY (1998) Antibodies to Von Willebrand factor-cleaving protease in acute thrombotic thrombocytopenic purpura. *N Engl J Med* 339: 1585–1594

44 Levy GG, Nichols WC, Lian EC, Foroud T, McClintick JN, McGee BM, Yang AY, Siemieniak DR, Stark KR, Gruppo R et al (2001) Mutations in a member of the ADAMTS gene family cause thrombotic thrombocytopenic purpura. *Nature* 413: 488–494

45 Zipfel PF, Jokiranta TS, Hellwage J, Koistinen V, Meri (1999) The factor H protein family. *Immunopharmacology* 42: 53–60

46 Schouten JP, McElgunn CJ, Waaijer R, Zwijnenburg D, Diepvens F, Pals G (2002) Relative quantification of 40 nucleic acid sequences by multiplex ligation-dependent probe amplification. *Nucleic Acids Res* 30: e57

47 Olie KH, Florquin S, Groothoff JW, Verlaak R, Strain L, Goodship TH, Weening JJ, Davin JC (2004) Atypical relapse of hemolytic uremic syndrome after transplantation. *Pediatr Nephrol* 19: 1173–1176

48 Donne RL, Abbs I, Barany P, Elinder CG, Little M, Conlon P, Goodship THJ (2002) Recurrence of hemolytic uremic syndrome after live related renal transplantation associated with subsequent de novo disease in the donor. *Am J Kidney Dis* 40: E22

49 Fortin MC, Schurch W, Cardinal H, Hebert MJ (2004) Complement factor H deficiency in acute allograft glomerulopathy and post-transplant hemolytic uremic syndrome. *Am J Transplant* 4: 270–273

50 Warwicker P, Donne RL, Goodship JA, Goodship THJ, Howie AJ, Kumararatne DS, Thompson RA, Taylor CM (1999) Familial relapsing haemolytic uraemic syndrome and complement factor H deficiency. *Nephrol Dial Transplant* 14: 1229–1233

51 Roodhooft AM, McLean RH, Elst E, Van Acker KJ, Vanacker KJ (1990) Recurrent hemolytic uremic syndrome and acquired hypomorphic variant of the third component of complement. *Pediatr Nephrol* 4: 597–599

52 Dragon-Durey MA, Fremeaux-Bacchi V, Loirat C, Blouin J, Niaudet P, Deschenes G, Coppo P, Herman FW, Weiss L (2004) Heterozygous and homozygous factor H deficiencies associated with hemolytic uremic syndrome or membranoproliferative

glomerulonephritis: report and genetic analysis of 16 cases. *J Am Soc Nephrol* 15: 787–795

53 Caprioli J, Bettinaglio P, Zipfel PF, Amadei B, Daina E, Gamba S, Skerka C, Marziliano N, Remuzzi G, Noris M (2001) The molecular basis of familial hemolytic uremic syndrome: Mutation analysis of factor H gene reveals a hot spot in short consensus repeat 20. *J Am Soc Nephrol* 12: 297–307

54 Goodship THJ, Liszewski MK, Kemp EJ, Richards A, Atkinson JP (2004) Mutations in CD46, a complement regulatory protein, predispose to atypical HUS. *Trends Mol Med* 10: 226–231

Towards a new classification of hemolytic uremic syndrome

Maren Salzmann[1], Michael Hoffmann[2], Gisa Schluh[1], Peter Riegler[3], Markus Cybulla[1], Hartmut P.H. Neumann[1]

[1]Department of Nephrology, Albert-Ludwigs-University, Hugstetter Straße 55, 79106 Freiburg, Germany; [2]Department of Laboratory Medicine, Albert-Ludwigs-University, Hugstetter Str. 55, 79106 Freiburg, Germany; [3]Department of Nephrology, General Hospital, Lorenz-Böhler-Str. 5, 39100 Bolzano, Italy

Introduction

Hemolytic uremic syndrome (HUS) is worldwide an important group of renal disorders. The criteria for the diagnosis of HUS are Coombs test-negative hemolytic anemia (with fragmented red cells), thrombocytopenia (thrombotic microangiopathy, TMA) and impairment of renal function, and a high risk for end-stage renal failure. The clinical features of HUS and thrombotic thrombocytopenic purpura (TTP) overlap: in general, nephrologists refer to the syndrome as HUS, hematologists as TTP. HUS can be classified by clinical and histopathological characteristics: infection, drug- or treatment-induced, genetic forms, pregnancy associated, acquired von Willebrand protease deficiency, superimposed on existing disorders or idiopathic. Genetical research has identified the involvement of three genes with multiple mutations, which allows a molecular classification of HUS and related disorders: factor H, membrane cofactor protein (MCP/CD46) and ADAMTS13. Current understanding is that the genetic status is only a predisposition and a "second event" is needed for manifestation of disease.

Clinical features

Clinical presentation

HUS presents with non-immunogenic anemia, low platelet count and acute renal insufficiency. Laboratory abnormalities include elevated lactate dehydrogenase, decreased haptoglobin, fragmentocytes in blood smear and negative Coombs test. In kidney biopsy specimens, TMA is evident by endothelial damage, glomerular microthrombi, necrosis and infiltrates of leukocytes. Renal impairment is mostly rapidly progressive, leading to end-stage renal failure within days or few weeks. There is an overlap of HUS with TTP, which is dominated by central nervous system symptoms.

Complement and Kidney Disease, edited by Peter F. Zipfel

Table 1 - A classification of HUS, TTP, and related disorders [6]

Infection
Shiga toxin and verocytotoxin-/Shiga-like toxin-producing organism; i.e. *Shigella dysenteriae* type 1 and enterohemorrhagic *E. coli*
Neuraminidase-producing organisms; red cell polyagglutination syndrome; i.e. *Streptococcus pneumoniae*
Human immunodeficiency virus
Drug- and treatment-induced
Oral contraceptives
Mitomycin C, 5-Fluorouracil, Gemcitabine, Bleomycin, Deoxycoformycin
Cyclosporin A, Tacrolimus, Sirolimus
Quinine
Ticlopidine, Clopidogrel
Interferon-α
"Crack" Cocaine
Genetic forms
Hemolytic complement abnormalities, e.g. Factor H
von Willebrand proteinase deficiency
Cobalamin metabolic disorders
Autosomal recessive not part of the above
Autosomal dominant not part of the above
Pregnancy-associated
Pregnancy-associated
Puerperium-associated
HELLP syndrome
Acquired von Willebrand protease deficiency
Auto-antibody against vWF protease
Superimposed on existing disorder
Associated with malignancy
Associated with bone marrow transplantation
Associated with vasculitis
Associated with glomerulonephritis
Accelerated phase hypertension
Currently unknown, idiopathic

Classical HUS

The classical or typical HUS occurs mainly in childhood. Most patients experience an episode of intestinal infection with diarrhea (D^+). Diarrhea-associated (D^+) HUS is caused by bacterial infections with production of Shiga-like verotoxins [1].

Of central importance are *E. coli* infections, mainly the 0157:H7 strain [2]. In India and Africa *Shigella dysenteriae* type 1 is a major cause [3]. Globatrioasylceramide (Gb3) receptor-mediated endothelial damage is predominantly found in the renal cortex [4], but Shiga-like toxins may also directly affect proximal tubular cells [5].

Clinical classification of HUS

A clinical classification aims towards etiology of HUS. Among several proposals, we prefer to use the suggested listing of Taylor and Nield [6] (Tab. 1). A major group is HUS associated with infections. In this group have to be mentioned also *Streptococcus pneumoniae*, Aeromonas and HIV. Another group is medication- und treatment-associated HUS. Noteworthy is that a number of immunosuppressive drugs used, among other instances, for prevention of kidney transplant rejection have been reported in context to HUS, such as cyclosporin A [7–10], tacrolimus [11], trapamycin [12, 13] azothioprine [14] and OKT 3 [15]. Some chemotherapeutics such as gemcitabine [16–21] and mitomycin [22, 23] have to be listed here. Other drugs are oral anticonceptiva, ticlopidin [24–28], clopidogrel [29] and quinine [30, 31]. Pregnancy-associated HUS may occur before childbirth or in the puerperium. Closely related are HELLP syndrome and HUS [32]. Associated diseases include systemic lupus erythematosus [33, 34], sclerodermia and Sjögren's syndrome, vasculitis and glomerulonephritis. Finally, malignancies and status after bone marrow transplantation [35] are well-known HUS-related conditions. With all these associations, one may wonder why only a minority of patients develop frank HUS but most do not. It can be speculated that these subjects may be carriers of a predisposition, and superimposition of such disorders, drugs etc. lead to clinical manifestation of HUS. Fascinating reports are those with familial occurrence of HUS. Both modes of transmission, autosomal dominant and autosomal recessive have been observed. Another interesting feature is relapsing HUS, which has been repeatedly found [36]. In a number of patients none of these conditions can be seen. For such patients the term idiopathic HUS is often used.

Pathogenesis

Complement alterations in HUS

Derangements of the complement system are known features in HUS. Namely, a decrease has been found of C3, Factor B and CH50 [37–39]. These data suggest an activation of the alternative complement activation. In 1994, Pichette et al. [40] reported a decrease of Factor H in familial HUS down to 5%. These findings were confirmed in 1999 by Noris [41]. Factor H is a central regulator protein of the alternative complement activation pathway. It regulates C3b by blocking Factor B and regulation of the C3bBb complex stability. Factor H also acts as a cofactor of Factor I for inactivation of C3b to iC3b [42, 43]. Factor H consists of 20 repetitive protein domains, so-called complement control protein modules (CCPs) or short consensus repeats (SCRs). The three-dimension structure is globular due to four highly conserved cysteines, and thus disulfide binding areas. The complement regulatory element has been attributed to CCP 1–4 [44–46]. Factor H has several complement regulatory functions including cell adhesion, receptor binding and interaction of cellular surfaces [47–50]. Other domains for C3 binding are located in CCP 6–15 and CCP 15–20 [51, 52]. Factor H is best characterized among the factor H protein family to which five Factor H-related proteins (FHR 1–5) and factor H-like protein (FHL1) also belong [53]. Deficiency of Factor H can be partial or complete. Absence of Factor H activity is associated with membranoproliferative glomerulonephritis type 2 [54]. This was found also in pigs [55] and in association with Factor H antibodies [56]. Another interesting regulator of the alternative pathway of complement activation is MCP, also called CD46. Together with Factor I, MCP degrades C3b and C4b bound to the cell surface [57]. MCP consists of four extracellular contiguous modules (SCRs) followed by a serine-threonine-proline-rich region, the transmembrane domain and a cytoplasmic tail.

Coagulation abnormalities

In contrast to global coagulation tests which reveal normal results in HUS and TTP, von Willebrand factor (vWF) is altered. It is produced by endothelial cells and stored there as well as in platelets. vWF is released after endothelium damage in very large multimers, but normally cleaved by a protease [58]. This process is enhanced by shear stress [59]. In TTP unusual large vWF multimers have been found in chronic relapsing and in familial disease [60]. A vWF-specific metalloproteinase cleaves vWF in the A2 domain between Tyr 1605 and Met 1606. Furlan [61] demonstrated, in non-familial TTP, an inhibitor of vWF cleaving protease and described a deficiency of the protease in familial TTP. The inhibitors are IgG antibodies [34]. Unusual large vWF multimers have not been reported in HUS. vWF cleaving protease activity can be measured by collagen binding affinity of degraded vWF [62], and thus HUS and TTP differentiated. The protease has been shown to be a member of the ADAMTS family (a disintegrin-like and metalloproteinase with thrombospondin motif) and is named ADAMTS13. It makes sense that TTP induced by vWF protease antibodies can effectively be treated by membrane plasmapheresis, a process which removes this antibody by substitution of antibody-free fresh frozen plasma.

Familial reports of HUS and TTP

Literature before 1998 comprises several reports of familial HUS including those from Bergstein [63], Farr [64], Edelsten and Tuck [65] and others [40, 66–72]. In most families, several siblings are affected, but the parents are not affected. These features have been interpreted mostly as autosomal recessive, although autosomal dominant transmission with incomplete penetrance is an alternative explanation. A clear autosomal dominant inheritance was given in the reports of Farr [64], Edelstein and Tucks [65] and Bergstein [63]. In these families age at onset was in adulthood, whereas in families with affected siblings only, age at onset varied from childhood to early adulthood. Recurrence of HUS after kidney transplantation was not rare [73, 74].

HUS-associated mutations

HUS associated with mutations of the FH1 gene

The major break through was successfully performed by the Newcastle-upon-Tyne group of Dr. Goodship [36]. Using genetic linkage studies in three HUS families, they identified the chromosomal locus at the long arm of chromosome 1, band 32 (1q32). The gene of Factor H is located in this area. Consequently, they analyzed the Factor H gene and found a point mutation, a missense mutation (Arg→Gly) in exon 20 (R1197G). The two other families did not show a FH1 mutation. They also analyzed a case with sporadic recurrent HUS, and found a deletion of four base pairs in exon 1. These findings gave evidence that sporadic HUS may apparently be a genetic disorder. In addition, astonishingly, serum Factor H levels were within normal range in the affected familial cases. FH1 consists of 23 exons encoding the FH1 protein with 1232 amino acids. So far extended studies have been reported from five centers from Newcastle (England) by Richards and Perkins [75, 76], from Bergamo (Italy) by Caprioli and Remuzzi [77, 78], from Madrid (Spain) by Sanchez-Corral [79], from Freiburg (Germany) by Neumann [80] and from Paris (France) by Dragon-Durey [81]. These series contain 7 cases from England, 4 from Italy, 4 from Spain, 17 from Germany and 10 from France. The 11 index cases are familial and 30 are sporadic cases affected by HUS. Ten relatives also showed the mutation of the index case in the given family but without disease. The total of the different mutations was 36 including 8 truncating mutations, 1 deletion of 1 amino acid and 27 missense mutations (Tab. 1). Interestingly, 22 of these mutations affect CCP 16–20, and another 7, located in CCP 1–15, produce proteins without CCP 16–20 (Fig. 1). C3 and Factor H levels of these cases are shown in Table 2. Of note, there is no clear correlation of C3 and Factor H, and many of the patients have normal C3 and normal or even elevated Factor H.

Table 2 - Mutations of the FH1 Gene associated with HUS

Mutation (cDNA Nucleotide)	Exon	Consequence (amino acid)	CCP	C3	Factor H	Family	Ref.
Cases with one mutation							
155-158 del 4 bp	2	Premature stop	1	low	low	sporadic	[36]
443 del (25bp)	3	Premature stop	2	low	low	sporadic	[81]
1271 C/A	9	400 Gln→Lys	7	normal	low	sporadic	[81]
1494 del A	11	Frameshift	8	low	normal	familial	[77]
1963 T/G	14	630 Cys→Trp	11	normal	normal	sporadic (carrier: father)	[80]
2091 G/A	14	673 Cys→Tyr	11	low	low	sporadic	[81]
2214 C/G	15	714 Ser→Stop	12	low	normal	sporadic (carrier: mother)	[80]
2373 ins A	16	Premature stop	13	low	low	sporadic	[81]
2621 G/A	17	850 Glu→Lys	14	high	high	sporadic	[80]
2751 A/G	18	893 His→Arg	15	low	low	sporadic	[81]
2770 T/A[a]	18	899 Tyr→Stop	15	low	low	familial	[81]
2816 T/A	18	915 Cys→Ser	15	low	low	sporadic	[81]
2846 C/T	18	924 Gln→Stop	15	low	low	sporadic	[81]
2940 C/T	19	956 Thr→Met	16	normal	high	sporadic (carrier mother)	[97]
3007 G/T	19	978 Trp→Cys	16	normal	normal	familial (carriers: mother, grandmother)	[80]
3200 T/C	20	1043 Cys→Arg	17	normal	normal	sporadic	[80]
3299 C/G	21	1076 Gln→Glu	18	normal	normal	sporadic (carrier father)	[80]
3429 A/G	22	1119 Asp→Gly	19	normal	normal	familial	[82]
3474 T/G	22	1134 Val→Gly	19	normal	high	sporadic (carrier: mother)	[80]
3497 T/G	22	1142 Tyr→Asp	19	low	high	sporadic	[80]
3542 T/C	22	1157 Trp→Arg	19	low	high	sporadic	[80]
3566 + 1 G/A	22	Splice defect	19	normal	high	sporadic	[80]

3620 C/T	23	1183 Trp→Arg	20	normal	normal	familial	[80]
3621 G/T	23	1183 Trp→Leu	20	low	high	sporadic	[97]
						n.r.	[78]
						familial	[81]
3624 C/G	23	1184 Thr→Arg	20	low	normal	sporadic	[82]
3639 T/G	23	1189 Leu→Arg	20	normal	high	familial	[97]
3654 G/A	23	1194 Gly→Asp	20	n.r.	n.r.	n.r.	[76]
3663 T/C	23	1197 Val→Ala	20	low	low	sporadic	[77]
3669 T/C	23	1199 Phe→Ser	20	low	low	sporadic	[81]
3701 C/T	23	1210 Arg→Cys	20	normal	normal	sporadic	[80]
3716 C/G	23	1215 Arg→Gly	20	normal	normal	sporadic	[73, 82]
3717 C/A	23	1215 Arg→Gln	20	low	normal	sporadic (carriers: father and grandfather)	[77]
3719 del ACA	23	1216 del Thr	20	low	normal	sporadic	[80]
3749 C/T	23	1226 Pro→Ser	20	normal	high	sporadic	[80]
3768-71 delAGAA	23	frameshift	20	low	normal	sporadic	[80]
del 24 bp[a]	23	Premature stop	20	low	low	sporadic	[77]
Cases with two mutations							
3135 A/T and	20	1021 Tyr→Phe	17	normal	normal	sporadic (carriers: father, mother)	[80]
3701 C/T	23	1210 Arg→Cys	20				
3299 C/G and	21	1076 Gln→Glu	18	n.r.	n.r.	sporadic	[82]
3559 del A	22	Frame shift	19				
3645 C/T[b] and	23	1191 Ser→Leu	20	n.r.	n.r.	familial	[82]
3663 T/C[b]	23	1197 Val→Ala	20				
3663 T/C[a] and	23	1197 Val→Ala	20	low	low	n.r.	[97]
Partial deletion[c]							

[a]*Homozygous;* [b]*Mutations on the same allele;* [c]*Other chromosome; n.r., not reported*

Mutations of CD46 (MCP) in HUS patients

MCP is a complement regulator (see above) and its gene is localized in a chromosomal area containing a cluster of complement-related genes. The gene is located in the same band as FH1, namely on chromosome 1, 1q32. The group in Newcastle-upon-Tyne described linkage of HUS to this area, but only in 1 of the used 3 families was a mutation of FH1 found. Therefore, in a new attempt, they tested 30 FH1-negative HUS families and detected MCP mutations in 3 families [82]. In one family, a heterozygote 6-base pair deletion causing deletion of two amino acids (D237, S238) was found. The other two families, one from Germany and one from Turkey showed the same mutation. The missense mutation 822 T/C (numbering from cDNA X59410) resulting in a serine to proline change was detected [82]. A relation of the two families was unknown and the DNA variant was not seen in 112 controls. In one family, the DNA variant was transmitted by the mother, and both patients showed a heterozygous mutation. In the other family, both patients showed the DNA variant on both alleles (homozygous mutation). MCP serum levels were reduced in the del DS 237-8 mutation carriers, but normal in the S206P carriers. The heterozygous S206P carriers showed 50% C3b binding, the homozygous cases complete lack of C3b binding. Noris et al. [83] reported, in a familial case of HUS, a deletion of 2 base pairs in codon 233, leading to a stop codon 236. This was also detected in the healthy mother. The expression of MCP protein was reduced by about 50% in all three mutation carriers (Tab. 3).

Mutations of the ADAMTS13 gene in TTP

ADAMTS13 plays a critical pathophysiological role for TTP (see above) [84]. Levy [85] performed a genome wide linkage scan in four families with familial or early onset of TTP. They identified an area on the long arm of chromosome 9, 9q34, as the candidate region of a susceptibility gene in a recessive model. After further narrowing the candidate area, they detected a protease or a protease cofactor candidate gene C90RF8, which exhibited homology to the ADAMTS family, now named ADAMTS13. This gene spans 29 exons, encompassing about 37 kb, encoding a protein of 1427 amino acids. Analyses of DNA sequences identified mutations within the ADAMTS13 gene in all four affected pedigrees studied. In addition, mutations were found in three TTP patients of other families. In total, they detected 12 different mutations, and found that affected individuals of one family had a homozygous mutation. Thus, in six families a homozygous or compound heterozygous ADAMTS13 mutation was shown, leaving one family (with the only splice site mutation) having only one heterozygous mutation. Further studies by Fujikawa [86] and Gerritsen et al. [87] confirmed the identity of ADAMTS13 as the vWF cleaving protease. Subsequently, nine other reports on ADAMTS13 mutations have been

Table 3 - Mutations of the MCP (CD46) gene associated with HUS

Index case	Homozygous/ heterozygous	Nucleotide	Exon	Consequence amino acid change	MCP activity	Age of onset (years)	Ref.
1	heterozygous	6-bp del (GACAGT)	6	loss of D237 + S 238	reduction of 50% in C3b binding	27; 31; 35[a]	[82]
2	heterozygous	T822 C trans	6	S206P change	reduction of 50% in C3b binding	8; 15[a]	[82]
3	homozygous	T822 C trans	6	S206P change	no detectable binding	9; 17[a]	[82]
4	heterozygous	2-bp A843-C844 del in codon 233	6	stop-codon at 236	reduction of around 50%	1.3; 9[a]	[83]

[a]*More than one patient in the family*

published [84, 88–96]. These are listed in Table 4. Summarizing these reports, nearly all cases show mutations of both alleles: the index cases of 8 families show homozygous mutations, and 22 families show compound heterozygous mutations. In 2 cases only one heterozygous mutation was found [85, 93]. Remarkable is the report of Antoine [84] on mutation analysis of the ADAMTS13 gene in two brothers affected by TTP and their parents. On the paternal allele in exon 2, they found the ADAMTS13 c. 130 C>T mutation, resulting in Q44X (Gln44stop). Two missense variants, c. 2195 C>T in exon 18 and c. 4006 C>T in exon 28 on the maternal allele, were not found in the 115 controls, leaving open whether one may represent a rare polymorphism and, if so, which. The brothers suffered from congenital deficiency of vWF cleaving protease (activity >3%), whereas the parents had about 50% activity and were asymptomatic. Where reported, ADAMTS13 activity was reduced lower than 7% in all patients.

Remarks to genetics and phenotype

Clinically, HUS and TTP show an overlap in a considerable number of patients. The pathogenesis of TTP currently appears to be better understood. Patients with TTP either harbor a genetic inactivation of both ADAMTS13 alleles of 9q34, and thus lack, or show minimal activity of, the specific ADAMTS13 proteinase. The other TTP patients obviously have an acquired disease with IgG antibodies against the proteinase. For HUS and FH1 gene mutations, only the Paris study has reported a decrease of Factor H by 50% in heterozygous and to lower than 5% in homozygous carriers. The latter all show membranoproliferative glomerulonephritis type 2 [81]. Difficult to understand are compound heterozygous carriers including the FH1 c. 3663 T/C mutation [75, 96]; similarly each of the mutations FH1 c. 3701 C/T and FH1 c. 3299 C/G have been observed in compound heterozygous and heterozygous carriers [75, 80]. The current understanding is that the genetic status is only a predisposition. A "second event" is needed for manifestation of the disease. In addition, 18 healthy mutation carriers have been detected in 11 families, suggesting a low lifelong risk for a "second event". This is in contrast to the risk for recurrent HUS after kidney transplantation, which seems to be about 50–80% in carriers and non-carriers of FH1 mutations [80], and about 60% in formally autosomal recessive HUS [97].

FH1 and MPC (CD46) are seemingly only two of several genes which contribute to the pathophysiology of HUS. At least 12 reported families with HUS are lacking a mutation in these genes [82, 83]. The importance of mutations in candidate genes can only be shown in a combined three-dimensional approach using molecular genetic DNA analysis and functional studies of the proteins as shown, but also using large national or international registries of clinically well-characterized index cases and relatives. Currently, there seems to exist only one such registry exceeding 100 cases [80].

Table 4 - Mutations of the ADAMTS13 Gene associated with TTP

Index case	Homozygous/ heterozygous	Nucleotide	Exon	Consequence amino acid change	Protease activity normal range (60–150%	Age of onset (years)	Ref.
1	homozygous	2074C/T	17	692 Arg→Cys	2–7%	nr	[85]
2	Compound	2851T/G	22	951Cys→Gly	2–7%	nr	[85]
	heterozygous	286C/G	3	96 His→Asp			
3	Compound	1582A/G	13	528 Arg→Gly	2–7%	nr	[85]
	heterozygous	3769-3770insT	27	Frameshift			
4	Compound	1193G/A	10	398 Arg→His	2–7%	nr	[85]
	heterozygous	3070T/G	24	1024 Cys→Gly			
5	Compound	304C/T	3	102 Arg→Cys	2–7%	nr	[85]
	heterozygous	587C/T	6	196 Thr→Ile			
6	Compound	2376-2401Δ	19	Frameshift	2–7%	nr	[85]
	heterozygous	3638G/A	26	C1213 Cys→Tyr			
7	heterozygous	1584+5G/A	13	Splice	2–7%	nr	[85]
8	Compound	4143 ins A	29	Frameshift	≤ 1 %	at birth	[88]
	heterozygous	3100A/T Nonsense	24	1034 Arg→Stop			
9	Compound	1523G/A	13	508 Cys→Tyr	nr	nr	[89]
	heterozygous	803G/C	7	268 Arg→Pro			
10	homozygous	1345C/T	12	449 Gln→stop	nr	nr	[89]
11	homozygous	TT deletion	15	Frameshift after His594	< 0.1 U/mL	At birth	[90]
12	Homozygous	932G/A	8	311 Cys→Tyr	decreased[a]	1.9	[91]
13	homozygous	703G/C	7	235 Asp→His	nr	1.8; 1	[91]
14	Compound	1058C/T	9	353 Pro→Leu	nr	1.7	[91]
	heterozygous	1370C/T	12	457 Pro→Leu			

Table 4 - continued

Index case	Homozygous/ heterozygous	Nucleotide	Exon	Consequence amino acid change	Protease activity normal range (60–150%	Age of onset (years)	Ref.
15	Compound	718-724Δ del G+C	7	frameshift	nr	0.3; 0.8 d	[91]
	heterozygous	2728C/T	21	910 Arg→stop			
16	homozygous	2279delG	19	Frameshift stop AA776	nr	0.6	[91]
17	homozygous	4143-4144insA	29	Frameshift stop AA1386	nr	1.5; 5.3	[91]
18	Compound	788C/G	7	263 Ser→Cys	< 2–5%	At birth	[92]
	heterozygous	4143insA	29	frameshift			
19	Compound	1058C/T	9	353 Pro→Leu	nr	2	[92]
	heterozygous	2728C/T	21	910 Arg→Stop			
20	Compound	4143insA	29	frameshift	< 2%	At birth	[92]
	heterozygous	3100A/T	24	1034 Arg→Stop			
21	homozygous	695T/A	7	232 Leu→Gln	2–6%	At birth	[92]
22	Compound	1169G/A	10	390 Trp→Stop	nr	At birth	[92]
	heterozygous	2549-2550delAT	20	frameshift			
23	Compound	1058C/T	9	353 Pro→Leu	< 2–6%	4	[92]
	heterozygous	4143insA	29	frameshift			
24	Compound	2728 C/T	21	910 Arg→Stop	< 2%	3	[92]
	heterozygous	4143insA	29	frameshift			
25	Compound	130C/T	18	732 Ala→Val	< 3%	nr	[84]
	heterozygous	2195C/T + 4006C/T[b]	28	1336 Arg→Trp			
26	heterozygous	414 + 1G/A	Intron 4	splice	< 3%	At birth	[93]
27	Compound	2017 A/T	17	1673 Ile→Phe	nr	At birth	[93]
	heterozygous	2723G/A	21	908 Cys→Tyr			

28	Compound heterozygous	577 C/T 1244 + 2T/G	6 Intron 10	193 Arg→Trp splice	nr	At birth	[93]
29	Compound heterozygous	686 + 1G/A 3367C/T	Intron 6 25	splice 1123 Arg→Cys	nr	At birth	[93]
30	Compound heterozygous	414 + 1G/A 2017A/T	Intron 4 17	splice 1673 Ile→Phe	< 3%	At birth	[93]
31	Compound heterozygous	A250V missense 5G>A	7 Intron 3	250 Ala→Val	< 3%	46	[94]
32	Compound heterozygous	1170 G/C 3735G/A	10 27	390 Trp→Cys 1245 Trp→stop	< 3%	10	[95]

[a]*Maybe due to streptococcal septicemia;* [b]*Unclear which of the 2 variants is a pathogenic mutation;* [c]*No mutation on the other allele found;* [d]*Two patients in the family; nr, not reported*

References

1 Calderwood S B, Acheson DWK, Keusch G T and Barrett T J (1996) Proposed new nomencalature for SLT (VT) family. *ASM News* 62: 118–119

2 Wong CS, Jelacic S, Habeeb RL, Watkins SL, Tarr PI (2000) The risk of the hemolytic-uremic syndrome after antibiotic treatment of *Escherichia coli* O157:H7 infections. *N Engl J Med* 342: 1930–1936

3 Bhimma R, Rollins NC, Coovadia HM, Adhikari M (1997) Post-dysenteric hemolytic uremic syndrome in children during an epidemic of *Shigella* dysentery in Kwazulu/Natal. *Pediatr Nephrol* 11:560–564

4 Louise CB, Obrig TG (1995) Specific interaction of *Escherichia coli* O157:H7-derived Shiga-like toxin II with human renal endothelial cells. *J Infect Dis* 172: 1397–1401

5 Hughes AK, Stricklett PK, Kohan DE (1998) Cytotoxic effect of Shiga toxin-1 on human proximal tubule cells.[see comment]. *Kidney Intern* 54: 426–437

6 Taylor CM, Neild GH (2005) Acute renal failure associated with microangiopathy (haemolytic-uremic syndrome and thrombotic thrombocytopenic purpura). In: AM Davison, JS Cameron, JP Grünfeld, C Ponticelli, E Ritz, CG Winearls, C van Ypersele (eds): *Oxford Textbook of Clinical Nephrology*. Oxford University Press, Oxford

7 Wiener Y, Nakhleh RE, Lee MW, Escobar FS, Venkat KK, Kupin WL, Mozes MF (1987) Prognostic factors and early resumption of cyclosporin A in renal allograft recipients with thrombotic microangiopathy and hemolytic uremic syndrome. *Clin Transplant* 11: 157–162

8 Neild GH, Reuben R, Hartley RB, Cameron JS (1985) Glomerular thrombi in renal allografts associated with cyclosporin treatment. *J Clin Pathol* 38: 253–258

9 Shulman H, Striker G, Deeg HJ, Kennedy M, Storb R, Thomas ED (1981) Nephrotoxicity of cyclosporin A after allogeneic marrow transplantation: glomerular thromboses and tubular injury. *N Engl J Med* 305: 1392–1395

10 Langer RM, Van Buren CT, Katz SM, Kahan BD (2002) *De novo* hemolytic uremic syndrome after kidney transplantation in patients treated with cyclosporine-sirolimus combination. *Transplantation* 73: 756–760

11 Lin CC, King KL, Chao YW, Yang AH, Chang CF, Yang WC (2003) Tacrolimus-associated hemolytic uremic syndrome: a case analysis. *J Nephrol* 16: 580–585

12 Heering P, Deppe CE, Farokhzad F, Helmchen U, Grabensee B (2002) Hemolytic uremic syndrome after renal transplantation: immunosuppressive therapy with rapamycin. *Nephron* 91: 177

13 Franco A, Hernandez D, Capdevilla L, Errasti P, Gonzalez M, Ruiz JC, Sanchez J; HUS-Sirolimus Spanish Study Group (2003) *De novo* hemolytic-uremic syndrome/thrombotic microangiopathy in renal transplant patients receiving calcineurin inhibitors: role of sirolimus. *Transplant Proc* 35: 1764–1766

14 Grino JM, Caralps A, Carreras L, Nogues R, Castelao AM, Romero R, Carreras M, Alsina J (1988) Apparent recurrence of hemolytic uremic syndrome in azathioprine-treated allograft recipients. *Nephron* 49: 301–304

15 Goodman DJ, Walker RG, Birchall IE, d'Apice AJ, Powell HR, Kincaid-Smith P (1991) Recurrent haemolytic uraemic syndrome in a transplant recipient on orthoclone (OKT 3). *Pediatr Nephrol* 5: 240–241

16 Fung MC, Storniolo AM, Nguyen B, Arning M, Brookfield W, Vigil J (1999) A review of hemolytic uremic syndrome in patients treated with gemcitabine therapy. *Cancer* 85: 2023–2032

17 Serke S, Riess H, Oettle H, Huhn D (1999) Elevated reticulocyte count--a clue to the diagnosis of haemolytic-uraemic syndrome (HUS) associated with gemcitabine therapy for metastatic duodenal papillary carcinoma: a case report. *Br J Cancer* 79: 1519–1521

18 Gross M, Hiesse C, Kriaa F, Goldwasser F (1999) Severe hemolytic uremic syndrome in an advanced ovarian cancer patient treated with carboplatin and gemcitabine. *Anti-Cancer Drugs* 10: 533–536

19 Muller S, Schutt P, Bojko P, Nowrousian MR, Hense J, Seeber S, Moritz T (2005) Hemolytic uremic syndrome following prolonged gemcitabine therapy: report of four cases from a single institution. *Ann Hematol* 84: 110–114

20 Venat-Bouvet L, Ly K, Szelag JC, Martin J, Labourey JL, Genet D, Tubiana-Mathieu N (2003) Thrombotic microangiopathy and digital necrosis: two unrecognized toxicities of gemcitabine. *Anti-Cancer Drugs* 14: 829–832

21 Walter RB, Joerger M, Pestalozzi BC (2002) Gemcitabine-associated hemolytic-uremic syndrome. *Am J Kidney Dis* 40: E16

22 Cattell V (1985) Mitomycin-induced hemolytic uremic kidney. An experimental model in the rat. *Am J Pathol* 121: 88–95

23 Verweij J, van der Burg ME, Pinedo HM (1987) Mitomycin C-induced hemolytic uremic syndrome. Six case reports and review of the literature on renal, pulmonary and cardiac side effects of the drug. *Radiother Oncol* 8: 33–41

24 Andersohn F, Hagmann FG, Garbe E (2004) Thrombotic thrombocytopenic purpura/haemolytic uraemic syndrome associated with clopidogrel: report of two new cases. *Heart* 90: e57

25 Manor SM, Guillory GS, Jain SP (2004) Clopidogrel-induced thrombotic thrombocytopenic purpura-hemolytic uremic syndrome after coronary artery stenting. *Pharmacotherapy* 24: 664–667

26 Moy B, Wang JC, Raffel GD, Marcoux JP (2000) 2nd Hemolytic uremic syndrome associated with clopidogrel: a case report. *Arch Intern Med* 160: 1370–1372

27 Chinnakotla S, Leone JP, Fidler ME, Hammeke MD, Tarantolo S (2000) Clopidogrel-associated thrombotic thrombocytopenic purpura/hemolytic uremic syndrome in a kidney/pancreas transplant recipient. *Transplantation* 70: 550–552

28 Evens AM, Kwaan HC, Kaufman DB, Bennett CL (2002) TTP/HUS occurring in a simultaneous pancreas/kidney transplant recipient after clopidogrel treatment: evidence of a nonimmunological etiology. *Transplantation* 74: 885–887

29 Bennett CL, Connors JM, Carwile JM, Moake JL, Bell WR, Tarantolo SR, McCarthy LJ, Sarode R, Hatfield AJ, Feldman MD et al (2000) Thrombotic thrombocytopenic purpura associated with clopidogrel. *N Engl J Med* 342: 1773–1777

30 Glynne P, Salama A, Chaudhry A, Swirsky D, Lightstone L (1999) Quinine-induced immune thrombocytopenic purpura followed by hemolytic uremic syndrome. *Am J Kidney Dis* 33: 133–137

31 Kojouri K, Vesely SK, George JN (2001) Quinine-associated thrombotic thrombocytopenic purpura-hemolytic uremic syndrome: frequency, clinical features, and long-term outcomes. *Ann Intern Med* 135: 1047–1051

32 McMinn JR, George JN (2001) Evaluation of women with clinically suspected thrombotic thrombocytopenic purpura-hemolytic uremic syndrome during pregnancy. *J Clin Apheresis* 16: 202–209

33 Kawasaki Y, Suzuki J, Nozawa R, Suzuki S, Suzuki H (2002) A 12-year-old girl with hemolytic uremic syndrome as initial symptom of systemic lupus erythematosus and a literature review. *Am J Nephrol* 22: 576–580

34 Tsai HM, Lian EC (1998) Antibodies to von Willebrand factor-cleaving protease in acute thrombotic thrombocytopenic purpura. *N Engl J Med* 339: 1585–1594

35 Rabinowe SN, Soiffer RJ, Tarbell NJ, Neuberg D, Freedman AS, Seifter J, Blake KW, Gribben JG, Anderson KC, Takvorian T et al (1991) Hemolytic-uremic syndrome following bone marrow transplantation in adults for hematologic malignancies. *Blood* 77: 1837–1844

36 Warwicker P, Goodship TH, Donne RL, Pirson Y, Nicholls A, Ward RM, Turnpenny P, Goodship JA (1998) Genetic studies into inherited and sporadic hemolytic uremic syndrome. *Kidney Int* 53: 836–844

37 Cameron JS, Vick R (1973) Letter: Plasma-C3 in haemolytic-uraemic syndrome and thrombotic thrombocytopenic purpura. *Lancet* 2: 975

38 Kaplan BS, Thomson PD, MacNab GM (1973) Letter: Serum-complement levels in haemolytic-uraemic syndrome. *Lancet* 2: 1505–1506

39 Monnens L, Molenaar J, Lambert PH, Proesmans W, van Munster P (1980)The complement system in hemolytic-uremic syndrome in childhood. *Clin Nephrol* 13: 168–171

40 Pichette V, Querin S, Schurch W, Brun G, Lehner-Netsch G, Delage JM (1994) Familial hemolytic-uremic syndrome and homozygous factor H deficiency. *Am J Kidney Dis* 24: 936–941

41 Noris M, Ruggenenti P, Perna A, Orisio S, Caprioli J, Skerka C, Vasile B, Zipfel PF, Remuzzi G (1999) Hypocomplementemia discloses genetic predisposition to hemolytic uremic syndrome and thrombotic thrombocytopenic purpura: role of factor H abnormalities. Italian Registry of Familial and Recurrent Hemolytic Uremic Syndrome/Thrombotic Thrombocytopenic Purpura. *J Am Soc Nephrol* 10: 281–293

42 Whaley K, Ruddy S (1976) Modulation of C3b hemolytic activity by a plasma protein distinct from C3b inactivator. *Science* 193: 1011–1013

43 Pangburn MK, Schreiber RD, Muller-Eberhard HJ (1977) Human complement C3b inactivator: isolation, characterization, and demonstration of an absolute requirement for the serum protein beta1H for cleavage of C3b and C4b in solution. *J Exp Med* 146: 257–270

44 Gordon DL, Kaufman RM, Blackmore TK, Kwong J, Lublin DM (1995) Identification of complement regulatory domains in human factor H. *J Immunol* 155: 348–356
45 Kuhn S, Skerka C, Zipfel PF (1995) Mapping of the complement regulatory domains in the human factor H-like protein 1 and in factor H1. *J Immunol* 155: 5663–5670
46 Kuhn S, Zipfel PF (1996) Mapping of the domains required for decay acceleration activity of the human factor H-like protein 1 and factor H. *Eur J Immunol* 26: 2383–2387
47 Hellwage J, Kuhn S, Zipfel PF (1997) The human complement regulatory factor-H-like protein 1, which represents a truncated form of factor H, displays cell-attachment activity. *Biochem J* 326: 321–327
48 DiScipio RG (1992) Ultrastructures and interactions of complement factors H and I. *J Immunol* 149: 2592–2599
49 Blackmore TK, Sadlon TA, Ward HM, Lublin DM, Gordon DL (1996) Identification of a heparin binding domain in the seventh short consensus repeat of complement factor H. *J Immunol* 157: 5422–5427
50 Blackmore TK, Hellwage J, Sadlon TA, Higgs N, Zipfel PF, Ward HM, Gordon DL (1998) Identification of the second heparin-binding domain in human complement factor H. *J Immunol* 160: 3342–3348
51 Sharma AK, Pangburn MK (1996) Identification of three physically and functionally distinct binding sites for C3b in human complement factor H by deletion mutagenesis. *Proc Natl Acad Sci USA* 93: 10996–11001
52 Zipfel PF (2001) Complement factor H: physiology and pathophysiology. *Semin Thromb Hemost* 27: 191–199
53 Zipfel PF, Skerka C (1994) Complement factor H and related proteins: an expanding family of complement-regulatory proteins? *Immunol Today* 15: 121–126
54 Ault BH, Schmidt BZ, Fowler NL, Kashtan CE, Ahmed AE, Vogt BA, Colten HR (1997) Human factor H deficiency. Mutations in framework cysteine residues and block in H protein secretion and intracellular catabolism. *J Biol Chem* 272: 25168–25175
55 Jansen JH, Hogasen K, Harboe M, Hovig T (1998) *In situ* complement activation in porcine membranoproliferative glomerulonephritis type II. *Kidney Int* 53: 331–349
56 Meri S, Koistinen V, Miettinen A, Tornroth T, Seppala IJ (1992) Activation of the alternative pathway of complement by monoclonal lambda light chains in membranoproliferative glomerulonephritis. *J Exp Med* 175: 939–950
57 Liszewski MK, Leung M, Cui W, Subramanian VB, Parkinson J, Barlow PN, Manchester M, Atkinson JP (2000) Dissecting sites important for complement regulatory activity in membrane cofactor protein (MCP; CD46). *J Biol Chem* 275: 37692–37701
58 Dent JA, Berkowitz SD, Ware J, Kasper CK, Ruggeri ZM (1990) Identification of a cleavage site directing the immunochemical detection of molecular abnormalities in type IIA von Willebrand factor.[erratum appears in *Proc Natl Acad Sci USA* Dec; 87 (23):9508]. *Proc Natl Acad Sci USA* 87: 6306–6310
59 Tsai HM, Sussman II, Nagel RL (1994) Shear stress enhances the proteolysis of von Willebrand factor in normal plasma. *Blood* 83: 2171–2179
60 Moake JL, Rudy CK, Troll JH, Weinstein MJ, Colannino NM, Azocar J, Seder RH,

Hong SL, Deykin D (1982) Unusually large plasma factor VIII:von Willebrand factor multimers in chronic relapsing thrombotic thrombocytopenic purpura. *N Engl J Med* 307: 1432–1435

61 Furlan M, Robles R, Galbusera M, Remuzzi G, Kyrle PA, Brenner B, Krause M, Scharrer I, Aumann V, Mittler U, Solenthaler M, Lammle B (1998) von Willebrand factor-cleaving protease in thrombotic thrombocytopenic purpura and the hemolytic-uremic syndrome. *N Engl J Med* 339: 1578–1584

62 Gerritsen HE, Turecek PL, Schwarz HP, Lammle B, Furlan M (1999) Assay of von Willebrand factor (vWF)-cleaving protease based on decreased collagen binding affinity of degraded vWF: a tool for the diagnosis of thrombotic thrombocytopenic purpura (TTP). *Thromb Haemost* 82: 1386–1389

63 Bergstein JM, Riley M, Bang NU (1992) Role of plasminogen-activator inhibitor type 1 in the pathogenesis and outcome of the hemolytic uremic syndrome. *N Engl J Med* 327: 755–759

64 Farr MJ, Roberts S, Morley AR, Dewar DF, Uldall PR (1975) The haemolytic uraemic syndrome – a family study. *Q J Med* 44: 161–188

65 Edelsten AD, Tuck S (1978) Familial haemolytic uraemic syndrome. *Arch Dis Child* 53: 255–256

66 Thompson RA, Winterborn MH (1981) Hypocomplementaemia due to a genetic deficiency of beta 1H globulin. *Clin Exp Immunol* 46: 110–119

67 Merrill RH, Knupp CL, Jennette JC (1985) Familial thrombotic microangiopathy. *Q J Med* 57: 749–759

68 Pirson Y, Lefebvre C, Arnout C, van Ypersele de Strihou C (1987) Hemolytic uremic syndrome in three adult siblings: a familial study and evolution. *Clin Nephrol* 28: 250–255

69 Mattoo TK, Mahmood MA, al-Harbi MS, Mikail I (1989) Familial, recurrent hemolytic-uremic syndrome. *J Pediatr* 114: 814–816

70 Roodhooft AM, McLean RH, Elst E, Van Acker KJ (1990) Recurrent haemolytic uraemic syndrome and acquired hypomorphic variant of the third component of complement. *Pediatr Nephrol* 4: 597–599

71 Kaplan BS, Kaplan P (1992) Hemolytic uremic syndrome in families. In: BS Kaplan, RS Trompeter, JL Moake (eds): *Hemolytic uremic Syndrome and Thrombotic Thrombocytopenic Purpura.* Marcel Dekker Inc., New York, 213–225

72 Kaplan BS, Leonard MB (2000) Autosomal dominant hemolytic uremic syndrome: variable phenotypes and transplant results. *Pediatr Nephrol* 14: 464–468

73 Kaplan BS, Chesney RW, Drummond KN (1975) Hemolytic uremic syndrome in families. *N Engl J Med* 292: 1090–1093

74 Kaplan BS, Papadimitriou M, Brezin JH, Tomlanovich SJ, Zulkharnain (1997) Renal transplantation in adults with autosomal recessive inheritance of hemolytic uremic syndrome. *Am J Kidney Dis* 30: 760–765

75 Richards A, Buddles MR, Donne RL, Kaplan BS, Kirk E, Venning MC, Tielemans CL, Goodship JA, GoodshipTH (2001) Factor H mutations in hemolytic uremic syndrome

cluster in exons 18-20, a domain important for host cell recognition. *Am J Human Genet* 68: 485–490

76 Perkins SJ, Goodship TH (2002) Molecular modelling of the C-terminal domains of factor H of human complement: a correlation between haemolytic uraemic syndrome and a predicted heparin binding site. *J Mol Biol* 316: 217–224

77 Caprioli J, Bettinaglio P, Zipfel PF, Amadei B, Daina E, Gamba S, Skerka C, Marziliano N, Remuzzi G, Noris M and Italian Registry of Familial and Recurrent HUS/TTP (2001) The molecular basis of familial hemolytic uremic syndrome: mutation analysis of factor H gene reveals a hot spot in short consensus repeat 20. *J Am Soc Nephrol* 12: 297–307

78 Remuzzi G, Ruggenenti P, Codazzi D, Noris M, Caprioli J, Locatelli G, Gridelli B (2002) Combined kidney and liver transplantation for familial haemolytic uraemic syndrome. *Lancet* 359: 1671–1672

79 Sanchez-Corral P, Perez-Caballero D, Huarte O, Simckes AM, Goicoechea E, Lopez-Trascasa M, de Cordoba SR (2002) Structural and functional characterization of factor H mutations associated with atypical hemolytic uremic syndrome. Am J Human Genet 71: 1285–1295

80 Neumann HP, Salzmann M, Bohnert-Iwan B, Mannuelian T, Skerka C, Lenk D, Bender BU, Cybulla M, Riegler P, Konigsrainer A et al (2003) Haemolytic uraemic syndrome and mutations of the factor H gene: a registry-based study of German speaking countries. *J Med Genet* 40: 676–681

81 Dragon-Durey MA, Loirat C, Cloarec S, Macher MA, Blouin J, Nivet H, Weiss L, Fridman WH, Fremeaux-Bacchi V (2005) Anti-factor H autoantibodies associated with atypical hemolytic uremic syndrome. *J Am Soc Nephrol* 16: 555–563

82 Richards A, Kemp EJ, Liszewski MK, Goodship JA, Lampe AK, Decorte R, Muslumanoglu MH, Kavukcu S, Filler G, Pirson Y et al (2003) Mutations in human complement regulator, membrane cofactor protein (CD46), predispose to development of familial hemolytic uremic syndrome. *Proc Natl Acad Sci USA* 100: 12966–12971

83 Noris M, Brioschi S, Caprioli J, Todeschini M, Bresin E, Porrati F, Gamba S, Remuzzi G and International Registry of Recurrent and Familial (2003) Familial haemolytic uraemic syndrome and an MCP mutation. *Lancet* 362: 1542–1547

84 Antoine G, Zimmermann K, Plaimauer B, Grillowitzer M, Studt JD, Lammle B, Scheiflinger F (2003) ADAMTS13 gene defects in two brothers with constitutional thrombotic thrombocytopenic purpura and normalization of von Willebrand factor-cleaving protease activity by recombinant human ADAMTS13. *Br J Haematol* 120: 821–824

85 Levy GG, Nichols WC, Lian EC, Foroud T, McClintick JN, McGee BM, Yang AY, Siemieniak DR, Stark KR, Gruppo R et al (2001) Mutations in a member of the ADAMTS gene family cause thrombotic thrombocytopenic purpura. *Nature* 413: 488–494

86 Fujikawa K, Suzuki H, McMullen B, Chung D (2001) Purification of human von Willebrand factor-cleaving protease and its identification as a new member of the metalloproteinase family. *Blood* 98: 1662–1666

87 Gerritsen HE, Robles R, Lammle B, Furlan M (2001) Partial amino acid sequence of purified von Willebrand factor-cleaving protease. *Blood* 98: 1654–1661

88 Kentouche K, Budde U, Furlan M, Scharfe V, Schneppenheim R, Zintl F (2002) Remission of thrombotic thrombocytopenic purpura in a patient with compound heterozygous deficiency of von Willebrand factor-cleaving protease by infusion of solvent/detergent plasma.[see comment]. *Acta Paediatrica* 91: 1056–1059

89 Kokame K, Matsumoto M, Soejima K, Yagi H, Ishizashi H, Funato M, Tamai H, Konno M, Kamide K, Kawano Y et al (2002) Mutations and common polymorphisms in ADAMTS13 gene responsible for von Willebrand factor-cleaving protease activity. *Proc Natl Acad of Sci USA* 99: 11902–11907

90 Savasan S, Lee SK, Ginsburg D, Tsai HM (2003) ADAMTS13 gene mutation in congenital thrombotic thrombocytopenic purpura with previously reported normal VWF cleaving protease activity. *Blood* 101: 4449–4451

91 Assink K, Schiphorst R, Allford S, Karpman D, Etzioni A, Brichard B, van de Kar N, Monnens L,van den Heuvel L (2003) Mutation analysis and clinical implications of von Willebrand factor-cleaving protease deficiency. *Kidney Int* 63: 1995–1999

92 Schneppenheim R, Budde U, Oyen F, Angerhaus D, Aumann V, Drewke E, Hassenpflug W, Haberle J, Kentouche K, Kohne E et al. (2003) von Willebrand factor cleaving protease and ADAMTS13 mutations in childhood TTP. *Blood* 101: 1845–1850

93 Matsumoto M, Kokame K, Soejima K, Miura M, Hayashi S, Fujii Y, Iwai A, Ito E, Tsuji Y, Takeda-Shitaka M et al (2004) Molecular characterization of ADAMTS13 gene mutations in Japanese patients with Upshaw-Schulman syndrome. *Blood* 103: 1305–1310

94 Uchida T, Wada H, Mizutani M, Iwashita M, Ishihara H, Shibano T, Suzuki M, Matsubara Y, Soejima K, Matsumoto M et al and Research Project on Genetics of Thrombosis (2004) Identification of novel mutations in ADAMTS13 in an adult patient with congenital thrombotic thrombocytopenic purpura. *Blood* 104: 2081–2083

95 Licht C, Stapenhorst L, Simon T, Budde U, Schneppenheim R, Hoppe B (2004) Two novel ADAMTS13 gene mutations in thrombotic thrombocytopenic purpura/hemolytic-uremic syndrome (TTP/HUS). *Kidney Int* 66: 955–958

96 Perez-Caballero D, Gonzalez-Rubio C, Gallardo ME, Vera M, Lopez-Trascasa M, Rodriguez de Cordoba S, Sanchez-Corral P (2001) Clustering of missense mutations in the C-terminal region of factor H in atypical hemolytic uremic syndrome. *Am J Hum Genet* 68: 478–484.

97 Loirat C, Niaudet P (2003) The risk of recurrence of hemolytic uremic syndrome after renal transplantation in children. *Pediatr Nephrol* 18: 1095–1101

Therapeutic strategies for atypical and recurrent hemolytic uremic syndromes (HUS)

Reinhard Würzner and Lothar B. Zimmerhackl

Departments of Pediatrics and Hygiene, Microbiology and Social Medicine, Innsbruck Medical University, Anichstr. 35, A–6020 Innsbruck, Austria

Introduction

In this review we use the term hemolytic uremic syndrome (HUS) for patients in whom the classical triad is present with hemolytic anemia (hemoglobin <10 mg/dl), thrombocytopenia (platelet count <150 000) and acute renal failure (serum creatinine above the 97th percentile and/or oliguria defined as urine output <0.5mL/kg/h). The term thrombotic microangiopathy is often used as a nomenclature combining the two entities, HUS and thromboctyopenic purpura [1]. However, since the pathogenetic pathways and the therapeutic options are divergent one should use this term with some caution.

The original description of HUS was given by Gasser, an internist, Gautier, a pediatrician, and other members of the University Hospital at Bern in 1955 [2]. Already in the original publication, Gasser did not talk about HUS but the plural: syndromes. Today both nomenclature and pathogenesis of HUS are not uniform. Several different diseases cause the same phenotype or clinical presentation, respectively. For clinical and practical purposes, the most frequently used differentiation is D^+ and D^-. In patients with D^+HUS, an association with Shiga toxin-induced gastroenteritis is found in over 90% of cases, usually as initial diarrhea (D^+, for diarrhea present) [3, 4]. Atypical HUS (aHUS) is thus often used as description for patients without prodromal diarrhea (D^-, for diarrhea absent). Among these diseases some have familial forms, some have recurrent disease, and others an "atypical" clinical presentation, whereby 'atypical' still comprises different causes. Severe neurological symptoms are sometimes considered as high-risk predisposing to long-term sequelae and death in the first episode [1, 3, 5, 6]. Positive family history or recurrent disease are also not associated with diarrhea and are therefore usually termed 'atypical'. These patients have a significantly worse prognosis. Renal sequelae in particular arterial hypertension and chronic renal failure are more frequent in atypical patients compared to the "classical HUS" (Fig. 1) [3, 6].

For this chapter, the therapeutic strategies are restricted to the description of aHUS and recurrent (D^-) HUS. A further narrowing is made by focusing on known

Complement and Kidney Disease, edited by Peter F. Zipfel

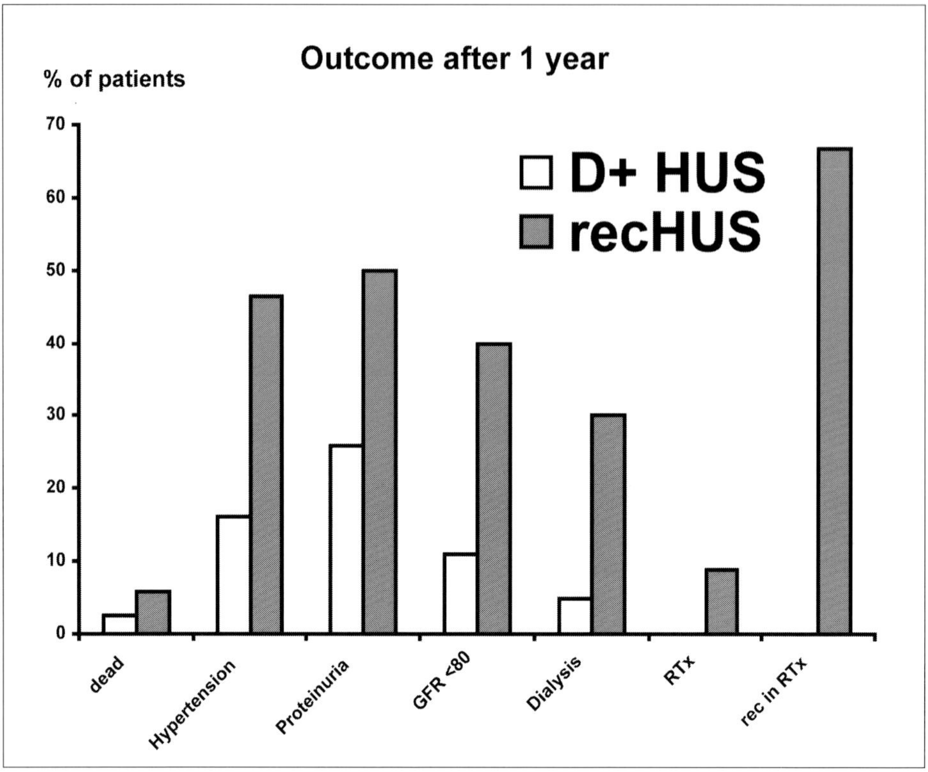

Figure 1
Outcome in patients with recurrent HUS
Data were obtained 1 year after initial presentation. RTx, renal transplantation; GFR, glomerular filtration rate; GFR in mL/min/1.73 m^2; rec, recurrent).

complement disorders. It should be emphasized that using that description only a portion of patients with HUS is characterized. According to recent data of the European research group for aHUS and related disorders, complement is involved in 50–70% of these patients (unpublished).

The disease has also been considered as thrombotic microangiopathy. Here the term thrombotic thrombocytopenic purpura (TTP, Morbus Moschcowitz) and HUS are considered as subgroups [1]. TTP, a disease preferentially of adults, and adult HUS have more in common and are often related to diseases of the von Willebrandt Factor (vWF), in particular to acquired diseases caused by antibodies to the vWF protease or inherited gene defects of the ADAMTS13 gene [5, 7]. In children, most patients have a gene defect and are symptomatic early in life. This subgroup of patients will not be in focus here.

Alternative hypothesis of Factor H involvement

Discussion of therapy concepts is difficult in the light of the unknown pathogeneses of HUS. Before these different options are discussed, we would like to present an alternative hypothesis as this may be relevant for the choice of therapy for at least a portion of affected patients.

Specific dysfunctional complement regulation predisposes to HUS, predominantly due to the lack of host cell protection from complement attack (see chapters by Skerka and Józsi, and Zipfel et al.). This is illustrated by the fact that a small fraction of Factor H gene mutations appear in the N-terminal part of the protein, which mediates the complement regulatory functions, but most of the mutations are clustered particularly in the C-terminal part of the molecule, which facilitates Factor H attachment, targeting Factor H action to the surface of the host cell (chapters by Noris and Remuzzi, and Zipfel et al.). Furthermore, lack of other cell-bound complement regulators also predisposes to aHUS (chapter by Goodship et al.).

The striking fact now is that by far most of the Factor H mutations in patients are heterozygous and are often associated with normal plasma levels, which is in sharp contrast to patients with dense-deposit disease (DDD, or membranoproliferative glomerulonephritis type II (MPGN type II)), who are usually homozygous deficient with consecutive absent or, in the case of a subtotal deficiency, very low levels [8, 9]. The most plausible explanation for this paradox is that the concentration of protective molecules is usually sufficient, but in certain (e.g., inflammatory) conditions higher concentrations would be required for protection, i.e., the defect predisposes rather than causes disease. However, normally nature allows heterozygosity for most molecules without predisposing to disease (and must allow this for the sake of preservation of species), provided the other half is (only!) absent and not interfering. This is in particular true for complement, as an important executor of innate immune defense [10, 11]. It is, therefore, somewhat strange to believe that as much as 50% of functionally active Factor H molecules, even in states of marked complement activation, should not appear to be sufficient.

We, therefore, propose that in some patients it is not the absence of 50% of functionally active molecules which predisposes to HUS, but rather the presence of dysfunctional molecules. A possible mechanism would be that the dysfunctional molecules attach to the host cell like their functionally active counterparts, not only lacking protection, but also preventing attachment of the active molecules as well. This hypothesis is corroborated by the fact that the quality of the mutations, and not the actual levels of the complement proteins, appear to be responsible for disease susceptibility and mortality rate, and that certain allotypes (not necessarily deficiency mutations!) are more often found predisposing to HUS [12, 13]. The resultant effect would even be more marked, if the affinity of the abnormal molecules is higher, and it would be interesting to test whether some of these aberrant molecules indeed possess a higher affinity for host cells [14]. This hypothesis should in particular be considered when plasma exchange (partially removing dysfunctional proteins) or plasma infusion (only slightly diluting these aberrant molecules) appear to be equal therapy options. As discussed below, the former option is often more successful, corroborating our hypothesis. Obviously, acquired autoimmune Factor H defects caused by autoaggressive antibodies, e.g., against Factor H itself [15, 16], will even more likely benefit more from plasma exchange, removing the autoaggressive particles, and/or immunosuppressive drugs, rather than plasma infusion alone.

Recommendations for therapy of recurrent/aHUS

In adult HUS/TTP, plasma exchange or plasma infusion have been recommended from the early 1990s onwards. At that time the pathophysiological rationale was (and is still) somewhat unclear, but survival in these patients dramatically improved with these regimen [10, 17]. In pediatric HUS, these recommendations were used also to improve outcome in severe HUS. Nevertheless, the effect of this procedure was not proven, but has been later copied for patients with recurrent HUS, for atypical and familial HUS [7, 17].

In adult TTP, plasma exchange is preferred to plasma infusion [7, 17, 18]. In adult HUS, the approach is similar. In pediatric HUS, no valid data are at present available and only anecdotal recommendations have been reported so far. For adult TTP, the plasma volume to be exchanged is generally estimated to be 40–60 mL/kg per session. Usually fresh frozen plasma (FFP) or FFP with 5% albumin have been used. Applying this regimen, the mortality dropped from >80% to below 10% in adults [7, 19].

For pediatric recurrent/atypical HUS, the recommendations are different. There is at present no differentiation for patients with coagulation disorders detectable by low C3 or C3d alone. As discussed before, different causes (mutations) may markedly affect the choice of therapy. The most relevant questions are:

(i) is there a difference between patients lacking Factor H and patients that have mutations in the Factor H gene with normal plasma concentrations?
(ii) should homozygous mutations be treated differently to heterozygous mutations?

These questions cannot be answered with the necessary evidence based on clinical studies, and we cannot differentiate the therapy at present. However, all these underlying mutations have to be considered in the future. Nowadays, the recommendations are based on more empirical approaches. Initially, the patient's clinical and especially his/her renal situation will influence the choice of therapy. Assuming the defect is causing increased complement turnover, usually determined by low C3 levels, but more accurately assessed by increased C3d or terminal complement complex (TCC) concentrations, the recommendations are plasma exchange usually 1–1.5 times the circulating plasma volume. In Figure 2 the therapies initiated in

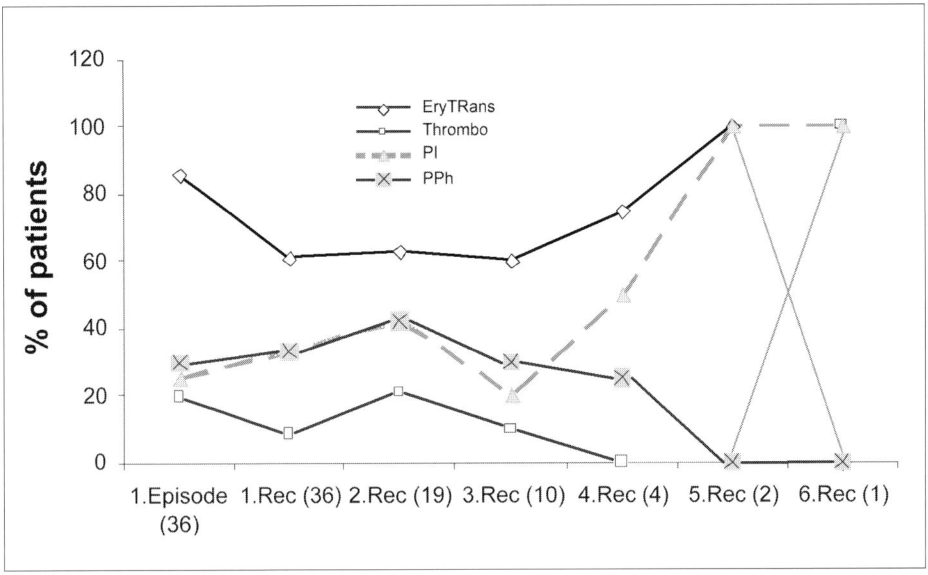

Figure 2
Present therapy of patients with recurrent HUS
Note: at first episode plasma exchange is used in 1/3 of patients (total number of patients in parentheses). In recurrent episodes (Rec) plasma infusion is the preferred therapy. PI, plasma infusion; PPh, plasmapheresis; EryTRans, transfusion of erythrocytes; Thrombo, transfusion of thrombocytes.

recurrent HUS from members of the European Research Group on aHUS are displayed. As noted, many colleagues start with plasma exchange and switch to plasma infusion in further recurrent episodes. The plasma volume used, assuming the above-mentioned approach, corresponds to 40–65 mL/kg body weight per exchange. Initially, the procedure should be performed daily or every other day until signs of hemolysis disappear: no fragmentocytes detectable (anymore), normalization of platelet counts, normalization of renal function [18, 20–23]. It should be noted that, under plasma therapy, the determination of LDH or haptoglobin as parameters of hemolysis may lead to false conclusions.

Prophylactic use of plasma either as exchange or as an infusion is, at present, even more empirical. Most pediatric nephrologists use 10–20 mL/kg per session between twice weekly to one infusion every 3 weeks. Some patients show particular signs of deterioration during periods of infections. Thus, one practical approach is to start with plasma exchange and try plasma infusion for prophylaxis. Most centers prefer a central line (Broviac, Permcath, etc.) or a *Cimino fistula* for long-term therapy.

Recommendations for mode of plasma therapy

At present there is no valid information available on whether plasmapheresis or plasma infusion are comparable modes of infusion. The rationale of giving plasma is that in some patients Factor H is missing or very low. In these situations plasma infusion is considered to be a replacement strategy [3, 23–25]. Assuming that there are no cryptic proteins present, infusion of plasma with normal Factor H concentrations should be a rationale treatment. For an assessment of the frequency of these infusions one has to know details on the half-life of Factor H.

Half-life of Factor H

Only few case reports on the half-life of Factor H have been published. In a patient with aHUS caused by complete Factor H deficiency, it was concluded that the half-life of Factor H is 6 days [24]. In a patient with HUS and heterozygous Factor H deficiency, we calculated a half-life of approximately 2.5 days (Fig. 3). However, in patients with heterozygous mutations with normal Factor H levels two different proteins are likely in the circulation [20, 26]. In the initial publication by Ault and co-workers [27], it was demonstrated that the mutated protein interfered with the regular secretion of normal Factor H. Thus, an uncertainty remains, and at present we do not know how the mutated protein is interfering with the C3b binding site and how this interacts with the inhibition of complement activation. Under the assumption that the mutated protein binds to the C3b binding site and is not acting as inhibitor, a therapy that removes the mutated protein would be favorable. In this case, plasmapheresis would be the more logical approach. Under the assumption that the mutated protein has no effect, plasma infusion should be sufficient.

Recommendations for plasma volume

The amount of plasma used and the route of application are variably discussed in the literature. In particular, it is not clear which volume is sufficient for therapy and even less is known on the amount necessary for prophylaxis. In this respect, Factor H levels may not really act as a surrogate marker for the reasons detailed above, but may nevertheless give a rough estimation. Using the above-mentioned half-lifes of Factor H of 2.5 and 6 days (in Fig. 3) the plasma volume substitutions can be calculated: A concentration of Factor H of lower than 200 mg/L would be reached between 70 and 130 h (4–7 days) after transfusion. Assuming no protein is present, our recommendation is to transfuse 60 mL/kg plasma per week. With this regimen a maximum of approximately 500 mg/L Factor H is reached shortly after transfusion and, following the decay, this calculation demands a regular transfusion of at least once weekly,

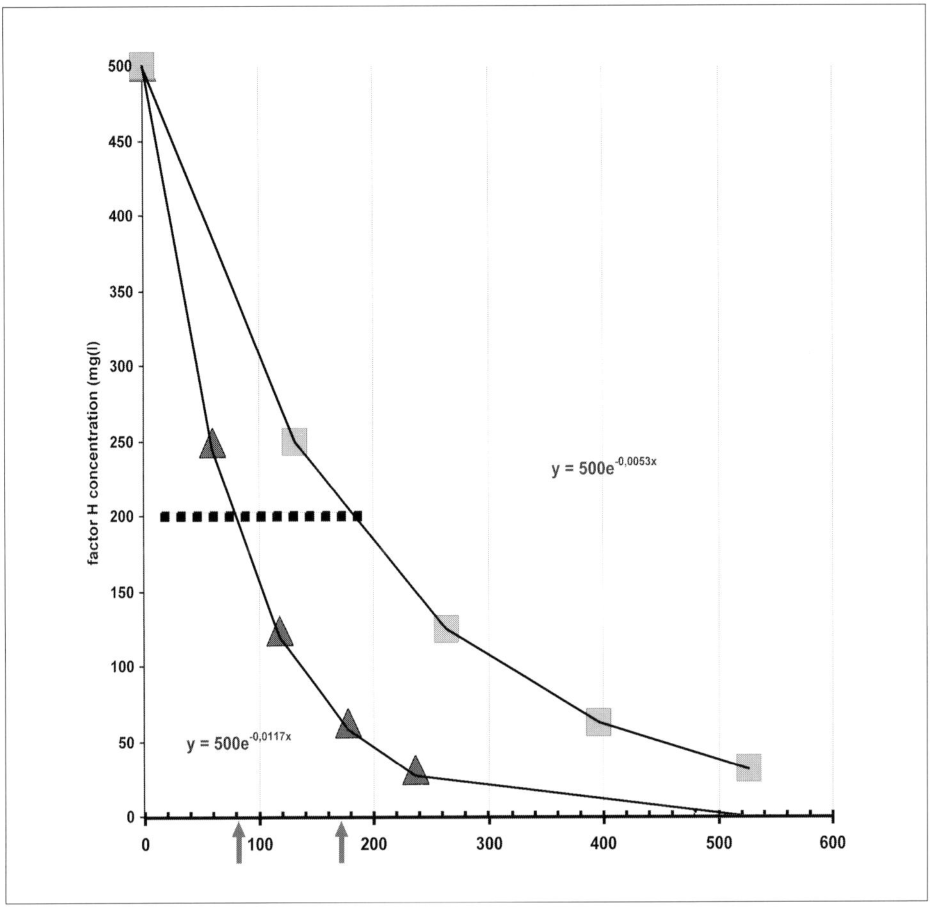

Figure 3
Half-life of Factor H
Half-life of Factor H in a patient with complete Factor H deficiency (quadrangle) and a patient with a heterozygous Factor H defect (triangle). Half-life was calculated from five plasmapheresis sessions from the decay to the next plasmapheresis. After plasmapheresis, factor H values were in the normal range. Half-life was calculated from the values using exponential regression. The Factor H concentration at 0 days was assumed to be 500 mg/L. Half-life was calculated as one compartment method using exponential regression (y=a.x, e^{-dt}).

but two transfusions of 25 mL/kg plasma per week may be as effective. However, all mathematical models will be different for each patient and, in fact, regular plasma infusions in 2-week intervals stopped the decline of kidney function in the homozygous deficient 2-year-old boy with the half-life of 6 days [24].

Choice of plasma

FFP, cryosupernatant, or cryoprecipitate may be used for transfusion [7, 17, 18]. Although these products will likely differ with regard to their Factor H contents, these differences in the various preparations are not known. Unfortunately, there are also no concentrations for Factor I reported. At present therefore, there is no evidence-based recommendation available for either one of the known preparations. FFP from units of whole blood or from plasmapheresis are comparable in view of therapeutic effectiveness. Plasma preparations may have several risks that have to be taken into account. In particular allergic reactions and anaphylaxis, transfusion-related acute lung injury and hemolysis to blood group antigens, especially A and B, have to be considered [18].

Risk of pathogen transmission due to plasma therapy

Risk of transmission of pathogens is small. The British Society for Haematology (BSH) estimates the risk of pathogen transmission for FFP to be in the range of 1 in 10 million for HIV, 0.2 in 10 million for hepatitis C, and 0.83 in 10 million for hepatitis B. Pathogen-reduced plasma (Octaplas®) has been proposed to be safer in this regard (see also below).

Patients likely to receive plasma products for long time should be considered for vaccination against hepatitis A and B. These patients with need for long-term plasma therapy may benefit from a reduced donor pool as used by Landau and co-workers [23] in their patients (preferentially single donor source), or pathogen-reduced FFP preparations. For children, the BSH recommends that only pathogen-reduced FFP (Octaplas®) should be used! If this demand cannot be met, the BSH recommends the donor to be at least male [18].

Risks of plasma infusion

There is considerable reservation regarding plasma infusion in children. In addition to the major risk factor, namely the transmission of infections, the chance of immunization of the patient is also of considerable importance. However, the patient's risk to generate HLA-related antibodies is not specifically known for patients with HUS. In general, the risk will increase with the number of plasma donors in the pool or the number of pools used consecutively in a patient. Since there is an imminent risk of developing end-stage renal disease with the need for kidney (and/or liver) transplantation in the future, any additional prior immunization should be avoided. Furthermore, infusion of plasma may increase the protein concentration in the patient's plasma and hence plasma viscosity. At present we are not aware of a published study on this topic. From our own experience and in particular from a personal communication (Dr. Guido Filler, Ontario, Canada), we conclude that the volume of plasma infusion should be limited. Filler reported that he has to perform plasma exchange once a month to maintain the patient's total protein concentration at least at 10 g/L. In addition to increased viscosity a high protein concentration causes plasma expansion and may therefore aggravate arterial hypertension. Since almost all patients have elevated blood pressure and more than a third have severe hypertension (> 30 mmHg above the 95th percentile of age) and require up to four different anti-hypertensive medications, this issue is of importance. This experience is comparable to the use of plasma in adult patients with TTP/HUS [7, 18, 19].

Risks of plasma exchange

Plasma exchange is a special dialysis technique using high blood flow vascular access. Thus, the risk of plasma exchange is comparable to side effects during renal replacement therapy. In general, during plasma exchange vascular reactions including shock symptoms are possible. The risk of shock is, however, considerably lowered by a tightly balanced volume exchange. In addition, the risk of bleeding or blood loss caused by the procedure itself has to be considered. Therefore, this therapy should be restricted to centers with experience with this therapy mode. Anaphylactic reactions are similar to those caused by plasma infusion [18].

Blood pressure control

As mentioned above, increased blood pressure is common in patients with recurrent/aHUS (see Fig. 1). In general, long time sequelae of increased blood pressure are a major problem in all children with renal disease. In particular, arteriosclerosis and myocardial infarction are responsible for death, morbidity or reduced quality of life in the long run. At present the aim should be to maintain arterial blood pressure in the normal range, preferentially below the 50th percentile. This should represent a supportive measure for all patients with recurrent/aHUS [1, 7, 28].

High viscosity

As mentioned above, plasma infusion may cause increased plasma viscosity. Increased viscosity aggravates arterial blood pressure and may induce thrombotic events. In these patients plasma exchange may be preferable to plasma infusion [7, 28, 29].

Coagulation

Coagulation disorders may interfere with the coagulation system. Despite the importance of this issue, the influence of the coagulation system in recurrent/aHUS is not well known. Future research should also focus on the interaction of complement disorders and coagulation.

Transplantation

Usually the differentiation between the two entities TTP and HUS is used regarding the recommendation for solid organ transplantation in recurrent/aHUS. Risk of recurrence of recurrent/aHUS is high and recommendations vary [28, 30, 31]. Indications for kidney transplantation according to Chiurchiu, Ruggenenti and Remuzzi are as as shown in Table 1 [21].

CD46-associated HUS

In addition to soluble factors, membrane-bound complement inhibitors, such as MCP (CD46) may be involved in HUS and renal failure. At present, only a few patients have been reported [32]. In theory, membrane-bound complement defects would cause cell damage preferentially in the organs that normally express this receptor protein. Since the kidney is a main organ expressing CD46, a renal transplantation should not be hampered by recurrence of the disease. However, CD46 is not restricted to the kidney and this may represent a residual risk of recurrence due to mechanisms yet to discover, when more patients with CD46 disorders are characterized [32].

Transplantation of liver and combined simultaneous liver and kidney transplantation

Under the assumption that Factor H is produced preferentially by the liver, attempts have been made to replace the normal Factor H by liver transplantation in patients with terminal renal failure. At present three patients are known. Although transplantation was initially successful, the long-term course was fatal; all three patients died. In these three patients unexplained coagulation problems were described. Therefore, this therapy option has not been performed by other groups to our knowledge. Is there a reason, why liver transplantation does not correct the Factor H disorder? Was the underlying deterioration forcing a liver transplantation so marked that this was the main reason for the fatal outcome? The pathogenesis is still

Table 1 - Indications for kidney transplantation and risk for recurrence of disease

Diagnosis	Recurrence risk	Kidney transplant
HUS typical (D^+)	0–10%	Recommended
HUS atypical	50–100%	Not recommended
HUS familial; recurrent or genetically determined	Almost 100%	Contraindicated

unclear. However, it should be considered that Factor H is not exclusively produced by the liver, but by other cells as well. Thus, under the assumption that some cryptic abnormal Factor H is still in the circulation, interference with complement inhibition is at least theoretically possible. As a consequence, inhibition of complement activation after transplantation may be affected even in patients with normal Factor H concentrations, similar to patients with heterozygous Factor H disorders. In these patients terminal renal insufficiency cannot be prevented, even by consequent plasma infusions, but possibly preserved (see chapter by Noris and Remuzzi)

Complement inhibitors

Several complement inhibitors have been solely tested *in vitro*, and only a few are available for therapy. Among the former, several recombinant proteins (including hybrids of two or more) of the family of regulators of complement activation (RCA), comprising CR1 (CD35), CR2 (CD21), MCP (CD46), DAF (CD55) and C4BP have been assessed *in vitro*, in addition to small molecule inhibitors, C3a- and C5a-receptor antagonists, and RNA aptamers [33, 34]. A very promising tool, however, would be a humanized version of an anti-factor B monoclonal antibody (mAb), specifically inhibiting the alternative pathway of complement [35]. Among the latter, especially soluble CR1 and anti-C5 antibodies are in clinical use. sCR1 (TP-10) has reached clinical studies. It has been tested in patients with acute respiratory distress syndrome, acute myocardial infarction and other diseases, but although sCR1 inhibited complement activation, there was no significant difference in the clinical outcome [36].

Anti-human C5 antibodies, a hallmark among anti-complement reagents, were first developed as mouse mAbs (N19-8), blocking TCC generation and C5a release to 97% and 70–80%, respectively, when added at an only one- to twofold molar excess over C5 [37]. The antibodies also worked in an *ex vivo* model of cardiopulmonary bypass [38]. Humanized versions of this and of a second human C5-blocking mAb (h5G1.1) have been successfully generated. That of the latter (scFv h5G1.1) has reached clinical trials as pexelizumab [39] and may represent an interesting therapy option for patients suffering from aHUS.

Conclusions

Therapy of recurrent/aHUS is not evidence based. However, several reports in the literature and the personal experience of the authors suggests that many patients with complement disorders respond well to any form of plasma therapy. Most centers prefer plasma exchange in the beginning and continue with plasma infusion as mainstay or prophylaxis. Despite this active and vigorous regimen many patients succumb to end-stage renal disease. Transplantation of either kidney alone or combined kidney/liver transplantation bear a high risk of recurrence. New therapeutic strategies are therefore absolutely mandatory. Hopefully, new insights in the mechanisms of disease and perhaps new complement inhibitors (such as humanized anti C5 antibodies) may improve treatment and outcome of these poor patients.

References

1 Zimmerhackl LB (1998) Epidemiology, pathogenesis and therapeutic modalities in hemolytic-uremic syndrome. *Kidney Blood Press Res* 21: 290–292

2 Gasser C, Gautier E, Steck A, Siebenmann Re, Oechslin R (1955) Hemolytic-uremic syndromes: bilateral necrosis of the renal cortex in acute acquired hemolytic anemia. *Schweiz Med Wochenschr* 85: 905–909

3 Gerber A, Karch H, Allerberger F, Verweyen HM, Zimmerhackl LB (2002) Clinical course and the role of shiga toxin-producing *Escherichia coli* infection in the hemolytic-uremic syndrome in pediatric patients, 1997–2000, in Germany and Austria: a prospective study. *J Infect Dis* 186: 493–500

4 Zimmerhackl LB (2000) *E. coli*, antibiotics, and the hemolytic-uremic syndrome. *N Engl J Med* 342: 1990–1991

5 Sutor AH, Thomas KB, Prufer FH, Grohmann A, Brandis M, Zimmerhackl LB (2001) Function of von Willebrand factor in children with diarrhea-associated hemolytic-uremic syndrome (D+ HUS). *Semin Thromb Hemost* 27: 287–292

6 Sautter S (2004) *Klinischer Verlauf und Veränderungen im Komplementsystem bei Kindern mit rekurrierendem Hämolytisch-urämischem Syndrom*. Doctoral thesis. Medical University Innsbruck

7 Ruggenenti P, Remuzzi G (1998) Pathophysiology and management of thrombotic microangiopathies. *J Nephrol* 11: 300–310

8 Dragon-Durey MA, Fremeaux-Bacchi V, Loirat C, Blouin J, Niaudet P, Deschenes G, Coppo P, Herman Fridman W, Weiss L (2004) Heterozygous and homozygous factor H

deficiencies associated with hemolytic uremic syndrome or membranoproliferative glomerulonephritis: report and genetic analysis of 16 cases. *J Am Soc Nephrol* 15: 787–795

9 Appel G, Cook TH, Hageman G, Jennette JC, Kashgarian M, Kirschfink M, Lambris J, Lanning L, Lutz HU, Meri S et al. (2005) Disease of the month: Membranoproliferative glomerulonephritis type II (dense deposit disease): An update. *J Am Soc Nephrol* 16: 1392–1403

10 Ross SC, Densen P (1984) Complement deficiency states and infection: epidemiology, pathogenesis and consequences of neisserial and other infections in an immune deficiency. *Medicine (Balt)* 63: 243–273

11 Würzner R, Orren A, Potter P, Morgan BP, Ponard D, Spath P, Brai M, Schulze M, Happe L, Götze O (1991) Functionally active complement proteins C6 and C7 detected in C6- and C7-deficient individuals. *Clin Exp Immunol* 83: 430–437

12 Caprioli J, Castelletti F, Bucchioni S, Bettinaglio P, Bresin E, Pianetti G, Gamba S, Brioschi S, Daina E, Remuzzi G, Noris M (2003) Complement factor H mutations and gene polymorphisms in haemolytic uraemic syndrome: the C-257T, the A2089G and the G2881T polymorphisms are strongly associated with the disease. *Hum Mol Genet* 12: 3385–3395

13 Esparza-Gordillo J, Goicoechea de Jorge E, Buil A, Carreras Berges L, Lopez-Trascasa M, Sanchez-Corral P, Rodriguez de Cordoba S (2005) Predisposition to atypical hemolytic uremic syndrome involves the concurrence of different susceptibility alleles in the regulators of complement activation gene cluster in 1q32. *Hum Mol Genet* 14: 703–712

14 Quigg RJ (2003) Complement and the kidney. *J Immunol* 171: 3319–3324

15 Dragon-Durey MA, Loirat C, Cloarec S, Macher MA, Blouin J, NIvet H, Weiss L, Fridman WH, Fremaux-Bacchi V (2005) Anti-Factor H Autoantibodies associated with atypical hemolytic uremic syndrome. *J Am Soc Nephrol* 16: 555–563

16 Dragon-Durey MA, Fremeaux-Bacchi V, Loirat C, Blouin J, Niaudet P, Deschenes G, Coppo P, Herman Fridman W, Weiss L (2004) Heterozygous and homozygous factor h deficiencies associated with hemolytic uremic syndrome or membranoproliferative glomerulonephritis: report and genetic analysis of 16 cases. *J Am Soc Nephrol* 15: 787–795

17 (2002) Guidelines on the diagnosis and management of the thrombotic microangiopathic hemolytic anemias. *Br J Haematol* 120: 556–573

18 O'Shaughnessy DF, Atterbury C, Bolton Maggs P, Murphy M, Thomas D, Yates S, Williamson LM (2004) British Committee for Standards in Haematology, Blood Transfusion Task Force. Guidelines for the use of fresh-frozen plasma, cryoprecipitate and cryosupernatant. *Br J Haematol* 126: 11–28

19 Rock A, Shumak KH, Buskard NA (1991) Comparison of plasma exchange with plasma infusion in the treatment of thrombotic thrombocytopenic purpura. Canadian Apheresis Study Group. *N Engl J Med* 325: 393–397

20 Caletti MG, Lejarraga H, Kelmansky D, Missoni M (2004) Two different therapeutic

regimes in patients with sequelae of hemolytic-uremic syndrome. *Pediatr Nephrol* 19: 1148–1152

21 Chiurchiu C, Ruggenenti P, Remuzzi G (2002) Thrombotic microangiopathy in renal transplantation. *Annals of Transplantation* 7: 28–33

22 Gerber A, Kirchhoff-Moradpour AH, Obieglo S, Brandis M, Kirschfink M, Zipfel PF, Goodship JA, Zimmerhackl LB (2003) Successful (?) therapy of hemolytic-uremic syndrome with factor H abnormality. *Pediatr Nephrol* 18: 952–955

23 Landau D. Shalev H, Levy-Finer G, Polonsky A, Segev Y, Katchko L (2001) Familial hemolytic uremic syndrome associated with complement factor H deficiency. *J Pediatr* 138: 412–417

24 Licht C, Weyersberg A, Heinen S, Stapenhorst L, Devenge J, Beck B, Waldherr R, Kirschfink M, Zipfel PF, Hoppe B (2005) Successful plasma therapy for atypical hemolytic uremic syndrome caused by factor H deficiency owing to a novel mutation in the complement cofactor protein domain 15. *Am J Kidney Dis* 45: 415–421

25 Nathanson S, Fremeaux-Bacchi V, Deschenes G (2001) Successful plasma therapy in hemolytic uremic syndrome with factor H deficiency. *Pediatr Nephrol* 16: 554–556

26 Manuelian T, Hellwage J, Meri S, Caprioli J, Noris M, Heinen S, Jozsi M, Neumann HP, Remuzzi G, Zipfel PF (2003) Mutations in factor H reduce binding affinity to C3b and heparin and surface attachment to endothelial cells in hemolytic uremic syndrome. *J Clin Invest* 111: 1181–1190

27 Ault BH, Schmidt BZ, Fowler NL, Kashtan CE, Ahmed AE, Vogt BA, Colten HR (1997) Human factor H deficiency Mutations in framework cysteine residues and block in H protein secretion and intracellular catabolism. *J Biol Chem* 272: 25168–25175

28 Loirat C, Niaudet P (2003) The risk of recurrence of hemolytic uremic syndrome after renal transplantation in children. *Pediatr Nephrol* 18: 1095–11101

29 Filler G, Radhakrishan S, Strain L, Hill Andrew, Knoll G, Goodship TH (2004) Challenges in the management of infantile factor H associated hemolytic uremic syndrome. *Pediatr Nephrol* 19: 908–911

30 Artz MA, Steenbergen EJ, Hoitsma AJ, Monnens LAH, Wetzels JFM (2003) Renal transplantation in patients with hemolytic uremic syndrome: high rate of recurrence and increased incidence of acute rejection. *Transplantation* 76: 821–826

31 Pirson Y, Lefebvre C, Lambert C, Warwicker P, Goodship T (2004) Familial hemolytic uremic syndrome: a privileged observation. *Bull Mem Acad R Med Belg* 159: 191–194

32 Richards A, Kemp EJ, Liszewski MK, Goodship JA, Lampe AK, Decorte R, Muslumanoglu MH, Kavukcu S, Filler G, Pirson Y et al (2003) Mutations in human complement regulator, membrane cofactor protein (CD46), predispose to development of familial hemolytic uremic syndrome. *Proc Natl Acad Sci USA* 100: 12966–12971

33 Mollnes TE (2004) Therapeutic manipulation in the complement system. In: J Szebeni (ed): *The Complement System. Novel roles in health and disease*. Kluwer, Norwell, Dordrecht, 483–516

34 Morgan BP, Harris CL (2003) Complement therapeutics; history and current progress. *Mol Immunol* 40: 159–170

35 Thurman JM, Lucia MS, Ljubanovic D, Holers VM (2005) Acute tubular necrosis is characterized by activation of the alternative pathway of complement. *Kidney Int* 67: 524–530

36 Zimmerman JL, Dellinger RP, Straube RC, Levin JL (2000) Phase I trial of the recombinant soluble complement receptor 1 in acute lung injury and acute respiratory distress syndrome. *Crit Care Med* 28: 3149–3154

37 Würzner R, Schulze M, Happe L, Franzke A, Bieber FA, Oppermann M, Götze O (1991) Inhibition of terminal complement complex formation and cell lysis by monoclonal antibodies. *Complement Inflamm* 8: 328–340

38 Rinder CS, Rinder HM, Smith BR, Fitch JC, Smith MJ, Tracey JB, Matis LA, Squinto SP, Rollins SA (1995) Blockade of C5a and C5b-9 generation inhibits leukocyte and platelet activation during extracorporeal circulation. *J Clin Invest* 96: 1564–1572

39 Fitch JC, Rollins S, Matis L, Alford B, Aranki S, Collard CD, Dewar M, Elefteriades J, Hines R, Kopf G et al (1999) Pharmacology and biological efficacy of a recombinant, humanized, single-chain antibody C5 complement inhibitor in patients undergoing coronary artery bypass graft surgery with cardiopulmonary bypass. *Circulation* 100: 2499–2506

Complement defects in children which result in kidney diseases: diagnosis and therapy

Christoph Licht and Bernd Hoppe

Children's Hospital of the University of Cologne, Pediatric Nephrology, Kerpener Str. 62, 50937 Cologne, Germany

Introduction

In recent years, several kidney diseases have been linked to defective complement regulation, namely to mutations of genes located within the "regulators of complement activation (RCA)" gene cluster on chromosome 1q32 [1]. First candidate genes in this region were Factor H (Factor H protein family) [2–5], and membrane cofactor protein (MCP, CD46) [6–8]. The clinical spectrum includes renal diseases like atypical hemolytic uremic syndrome (aHUS), membranoproliferative glomerulonephritis type II (MPGN II), isolated microscopic hematuria, collagen type III glomerulopathy, and IgA nephropathy [9]. However, diseases with a mainly extrarenal manifestation have also been linked to the complement system: systemic lupus erythematosus (SLE) [10, 11], increased susceptibility to meningococcal infections resulting in meningococcal meningitis [12–14], recurrent microbial infections and chronic inflammatory conditions such as rheumatoid arthritis [15, 16], and immune evasion of tumor cells [17, 18], as well as disorders in the coagulation system [19–21].

A key finding in patients with complement-based renal diseases is the activation of the alternative complement pathway, in the majority of cases caused by heterozygous or homozygous Factor H deficiency [9]. However, not only deficiency of Factor H but also mutations affecting protein function but not plasma availability are identified as causing disease (aHUS [22–25], MPGN II [26]). Dragon-Durey et al. [27] recently described the presence of anti-Factor H autoantibodies as a new pathomechanism of functional Factor H deficiency also causing aHUS. In addition, deficiency in membrane-anchored MCP and Factor I are now also known to cause aHUS [6–8, 28].

By contrast, the paroxysmal nocturnal hemoglobinuria (PNH) is not based on Factor H or MCP but on a defect of the decay-accelerating factor (DAF, CD55). DAF is a glycophosphatidylinositol-linked membrane protein expressed on endothelial cells and erythrocytes. A defect of this protein-lipid linkage results in the failure to express DAF and causes PNH. Unregulated complement activation on the surface

Complement and Kidney Disease, edited by Peter F. Zipfel

of erythrocytes causes recurrent intravascular hemolysis and – in the long run – leads to chronic hemolytic anemia and venous thrombosis [19, 29–32].

In principal, all of these diseases can be found in children. Thus, consequent screening for defects in the regulation of the complement system will allow the actual incidence/prevalence of the (complement-based subtypes of the) above-mentioned diseases to be identified. In the following chapter, however, we focus on aHUS and MPGN II, which are so far the most common known complement-based renal diseases in children [9].

The complement system

Introduction

The complement system is part of the innate immune system. It represents one of the major recognition systems of innate immunity and an important effector mechanism of humoral (antibody-mediated) adaptive immunity. The central physiological functions of the complement system are:
(i) Defense against pyogenic microbial infections
(ii) Bridging innate and adaptive immunity
(iii) Disposing of immune complexes and products of inflammatory injury.

The complement system consists of a family of multiple plasma and surface-bound proteins, which interact with one another and other proteins of the immune system in a tightly regulated fashion. Activated complement proteins (e.g., central complement component C3/C3b) bind, e.g., to the surface of microbes or to antibodies within antigen-antibody complexes and induce the activation of the late steps of the complement cascade, which are (1) opsonization and phagocytosis, (2) induction of inflammation, and (3) formation of the membrane attack complex (MAC) and cytolysis [29–31].

Microbes on which complement is activated become coated with C3b or C4b (opsonization), and are phagocytosed by the binding of these proteins to specific receptors on macrophages and neutrophils. C5a, C4a, and C3a, also called anaphylatoxins, are proteolytic fragments of C5, C4, and C3, which are generated when these complement components are activated. They induce acute inflammation by acting on mast cells and neutrophils. Finally, binding of C3b to microbes and activation of the late components of the complement cascade (C5b-C9, terminal complement complex, TCC) leads to the formation of the MAC, which causes osmotic lysis of the microbes [29–31].

With few exceptions, host cells are protected against activation of the complement system on their surface by cell membrane-anchored inhibitors. Microbial cells, by contrast, lack these inhibitors. Thus, the complement system is able to "differentiate" between "self" and "non-self", which guarantees the directed attack against invading microbial cells and prevents a generalized self attack with fatal consequences for the host [29–31].

Regulation of complement activation

Complement activation occurring on foreign cells such as invading microbes is part of a vital defense mechanism of the body, providing the possibility of immediate immune reaction with the final goal of attacking and eliminating these microbes. As a consequence, complement activation on "self cells" must be tightly regulated and controlled by both soluble, e.g., Factor H, and cell-membrane anchored, e.g., MCP, DAF, regulators to avoid fatal self attack [33–36].

Activation of the complement system occurs via one of the following pathways:

(i) Alternative pathway
(ii) Classical pathway
(iii) Lectin pathway.

The two main pathways of complement activation are the alternative pathway, which is – at low level – continuously activated with the early steps of the complement cascade occurring indiscriminately on any surface in the absence of antibody, and the classical pathway, which is activated by antigen-antibody complexes [29–31, 37]. The lectin pathway is activated by microbial polysaccharides and circulating lectin molecules and follows, except for the different induction, the classical pathway. Both the classical and the alternative pathway merge at the level of C3 resulting in the generation of C3b (Fig. 1) [29–31].

Under physiological conditions, there is constant low rate of C3 cleavage in plasma, generating C3b – a process called "C3 tick over". A small amount of C3b becomes covalently bound to the surfaces of cells, including host cells and microbes. Surface-attached C3b binds Factor B, which is then cleaved by Factor D to generate Factor Bb. Factor Bb remains attached to C3b, resulting in the alternative pathway C3 convertase C3bBb. Properdin binds to and stabilizes C3bBb. The function of C3bBb is to cleave more C3 molecules, thereby setting in motion the activation loop of C3 within the alternative pathway of the complement system [29–31].

On mammalian ("self") cells, C3bBb becomes rapidly degraded by several surface-attached and membrane-anchored regulatory proteins. On microbial cells, however, in the absence of such regulators, activation of the (alternative pathway of the) complement system proceeds. Some of the C3b molecules generated by the C3 convertase bind to C3bBb, thereby forming the alternative pathway C5 convertase C3bBbC3b [29–31].

Within the classical pathway, activation of C3 occurs through the action of the classical pathway C3 convertase: C1 binds to antigen-antibody complexes and

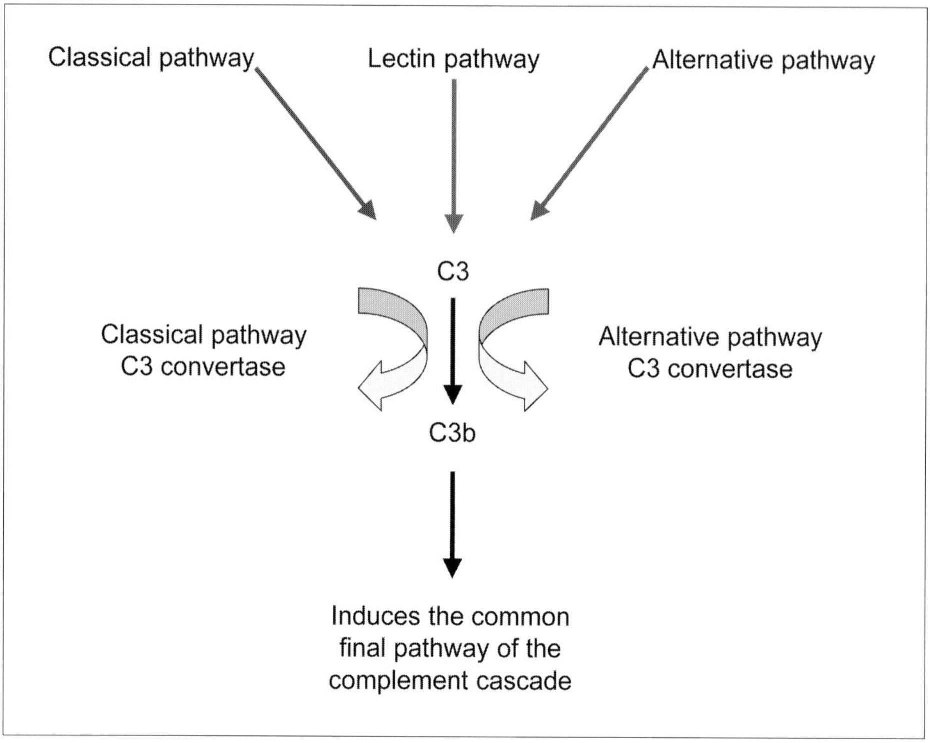

Figure 1
Key steps of the complement cascade
The complement system can be activated by the alternative, the classical or the lectin pathway,which merge in the activation of C3, which is followed by the common final steps of the complement cascade.

becomes activated. C4 binds to bound C1 and becomes activated by C1, resulting in the formation of C4b. C4b can also directly bind to the surface of microbes. Independent of its binding site, however, C4b then binds C2a – with C2 also being activated by C1 – to form C4bC2a, the classical pathway C3 convertase. Similar to the alternative pathway C3 convertase, the classical pathway C3 convertase also cleaves, and thereby activates, more C3 molecules and binds C3b, resulting in the formation of the classical pathway C5 convertase C4bC2aC3b [29–31].

Both the alternative and the classical pathway C5 convertase then activate C5, resulting in the formation of C5b. C5b activates the TCC, and leads to the formation of the MAC, consisting of the complement components C5b-C9 [29–31].

Activation of the complement cascade and the stability of active products of the complement system need to be tightly regulated to prevent complement activation on normal host cells, and to limit the duration of complement activation even on microbial cells and antigen-antibody complexes. Complement activation is regulated by a family of humoral or membrane-bound proteins summarized as RCA. Members of this family are C1 inhibitor (C1 INH), Factor I, Factor H, Factor H like protein-1 (FHL-1), C4 binding protein (C4BP), MCP, type 1 complement receptor (CR1, CD35), DAF, and CD59 [29–31].

C1 INH inhibits the proteolytic activity of C1. MCP, CR1 and DAF inhibit the binding of regulatory proteins to C3b and C4b. Furthermore, C3b is bound by Factor H (alternative pathway), and C4b is bound by C4BP (classical pathway). By binding to C3b and C4b, Factor H and C4BP competitively inhibit the binding of other components of the C3 convertase, such as Bb (alternative pathway) and C2a (classical pathway), thus inhibiting further activation of the complement cascade. C3b and C4b are proteolytically degraded by Factor I. MCP, Factor H, C4BP, and CR1 serve as cofactors for Factor I-mediated cleavage of C3b and C4b. CD59 inhibits the formation of the MAC [29–31].

The differential attachment and presence of these regulators on mammalian ("self") but not on microbial cells results in selective inhibition of the complement system on host cells and allows complement activation to proceed on microbial cells [29–31].

The complement system and disease

The complement system is involved in diseases in two general ways: deficiencies in any of the protein components of the complement cascade may lead to abnormal complement activation resulting in either deficient or unrestricted activation of the complement system (defect of regulatory components) with too much complement activation at the wrong time or the wrong site [29–31]. With respect to renal diseases, however, in most cases deficiency results in unrestricted activation of the complement cascade, especially of the alternative pathway [38].

As described above, there is a redundant system of factors, which are in charge of a fine-tuned regulation of the activation of the complement cascade, the RCA. The functional redundancy of these factors ensures that the activation of the complement cascade occurs specifically "at the right time and at the right site" and self attack and damage of self surfaces of the mammalian organism are avoided. Furthermore, the distribution of the proteins that regulate the activation of the complement system directs the action of the complement system specifically to the surface of the target cells, e.g., invading microbes. While, in general, microbial cells lack surface-bound regulators for the activation of the complement system, mammalian cells, i.e., host cells, are protected by the presence of such regulatory proteins (e.g., MCP, CR1 and DAF). In addition, a group of humoral factors (e.g., Factor H, FHL-1, Factor I) acts complementary to these cell membrane-bound proteins. These factors are primarily present as circulating plasma proteins, which bind and attach to host cell surfaces. The serine protease Factor I cleaves C3b and C4b using Factor H and other factors, e.g., MCP, C4BP, and CR1, as cofactors [29–31].

However, it has been shown that Factor H, to become functionally active, needs to bind to cell surfaces [21, 37, 39]. Factor H binds with high affinity to cell membranes rich in sialic acid, as found in mammalian cells different from microbial cells, which again directs the action of the complement system to attack invading microbes and to protect mammalian host cells [21, 37, 39]. When bound to a cell membrane, Factor H functions as cofactor of Factor I in mediating the cleavage of cell membrane-bound C3b and C4b. Furthermore, Factor H binds C3b, thereby preventing the binding of other factors to C3b, e.g., of Bb, which would result in the formation of the alternative pathway C3 convertase C3bBb – or even displacing already bound Bb from C3b, thereby dissociating and inactivating the alternative pathway C3 convertase [29–31].

A central step within the sequential events of complement activation is the activation of the third complement component – C3 – resulting in the generation of C3b. Responsible for the activation of C3, however, are the classical and/or the alternative pathway C3 convertases C3bBb and C4bC2a. All previously described regulatory mechanisms controlling the activation of the complement system are positioned around this center: they either cleave C3b or C4b themselves, or inhibit the formation of the C3 convertase complex or dissociate the already formed complex [29–31].

In addition, there is another humoral factor that is assumed to stabilize the (alternative pathway) C3 convertase, called C3 nephritic factor (C3NeF). C3NeF is supposed to be an IgG antibody, which covalently binds to C3bBb, thereby preventing Factor H-induced dissociation of this complex. With respect to the final result, the presence of C3NeF displays a comparable effect to any other situation in which deficiency or defect of the previously described regulatory proteins lead to unrestricted activation of the complement system (Fig. 2) [29–31, 40, 41].

In summary, each defect – either quantitative deficiency or functional defect – of one or more of the proteins forming this tight and redundant network of control of complement activation (alternative and classical pathway), or the presence of C3NeF (alternative pathway), unavoidably leads to unrestricted complement activation, which bears the potential of causing disease [29–31].

Kidney diseases as result of complement defects in children

Atypical hemolytic uremic syndrome

In 1952, Symmers [42] defined the disease thrombotic microangiopathy (TMA) as a "lesion of vessel wall thickening (mainly affecting arterioles and capillaries) with

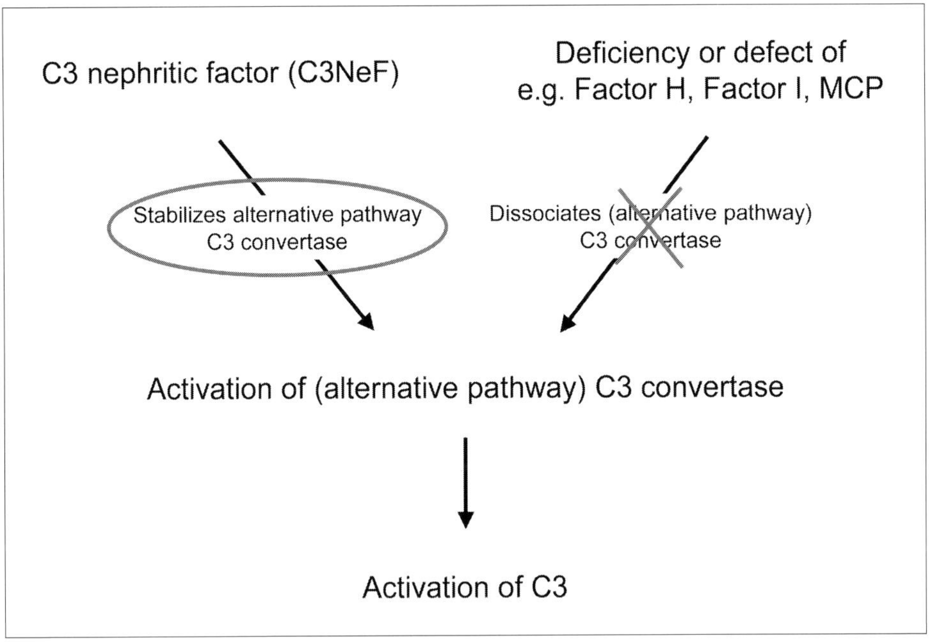

Figure 2
Unrestricted activation of (alternative pathway) C3 convertase
Both presence of C3NeF and deficiency or functional defect of Factor H, Factor I and MCP result in unrestricted activation of (alternative pathway) C3 convertase and subsequently of C3.

swelling or detachment of the endothelial cells from the basement membrane, accumulation of fluffy material in the subendothelial space, intraluminal platelet thrombi, and partial or complete obstruction of the vessel lumina". Depending on the distribution of the microvascular lesions and the age of the patient, different disease subtypes can develop: while in adults neurological symptoms, e.g., transient ischemic attacks (TIA), prevail as a sign of cerebral affection, acute renal failure represents the main clinical complication in children, indicating that the kidney is the preferred disease target in this age group [43]. The TMA subtype found in adults is called thrombotic thrombocytopenic purpura (TTP) or Morbus Moshcowitz, whereas the subtype found in children is called hemolytic uremic syndrome (HUS) [43–45].

HUS is defined by the triad microangiopathic hemolytic anemia with the appearance of fragmentocytes/schistocytes and thrombopenia followed by acute renal failure (uremia), which was first described by Gasser et al. in 1955 [44]. HUS occurs with a frequency of 5–10/100,000 children and represents the major cause of acute renal failure in childhood [46]. There are two main forms of HUS: typical D^{+}HUS or Shiga toxin HUS (Stx HUS), and atypical D^{-}HUS. Stx HUS usually follows an episode of gastroenteritis caused by verocytotoxin- or Shiga (like) toxin-producing enterohemorrhagic *Escherichia coli* (EHEC), mostly of the serotype O157:H7 [46]. It affects male and female patients with the same frequency. Relapses are extremely rare, and compared to aHUS, prognosis is usually good with end-stage renal failure (ESRF) in <10% [46]. Atypical D^{-}HUS, however, can occur in a sporadic or familial (inherited) fashion, is characterized by the absence of antecedent diarrhea, the tendency to pre- and post-transplant relapse (about 50% after renal transplantation in children) and often a poor outcome with development of ESRD or death. Atypical D^{-}HUS accounts for 5–10% of all HUS cases and prefers male patients (male:female, 2:1) [46–50].

Pathogenesis of typical and atypical HUS merge in a common final pathway: damage of blood vessel endothelial cells causes exposure of the basement membrane, which activates both the complement and the coagulation systems. Activation of the coagulation system causes formation of platelet-rich thrombi, which are carried with the bloodstream and later occlude peripheral small vessels (Fig. 3A and B) [42, 46]. In children, this process mainly affects the kidneys and causes transient or permanent loss of organ function [42, 46]. It is still unclear why childhood HUS, both the typical and the atypical type of the disease, mainly affects the kidneys before other organ systems. A possible explanation for Stx HUS is the high expression level of Gb3 receptors in endothelial cells of renal vessels, resulting in an increased Stx binding and thereby higher susceptibility to Stx-mediated damage [51, 52].

In recent years, there has been a remarkable increase in the understanding of the molecular pathomechanisms responsible for TMA like TTP (Morbus Moshcowitz) and aHUS [42, 46, 53]. For aHUS, defective control of the activation of the alternative complement pathway by Factor H deficiency caused by Factor H gene mutations or autoantibodies against Factor H [34, 37, 46, 54–59], or by MCP deficiency caused by MCP gene mutations [6–8] has been identified as key pathomechanism.

Before deficiency or functional defects of Factor H (or other complement regulators) were proven to be crucial for the development of aHUS, hypocomple-

---

Figure 3
HUS
Renal biopsy with typical glomerular changes of HUS including segmental thickening of glomerular basement membranes with (A) double contours, congestion of capillaries and increase in extracellular matrix (Masson trichrome stain, original magnification × 100) and (B) thrombemboli occluding the lumen of the glomerular arterioles (EM by Rüdiger Waldherr, Heidelberg, Germany).

A

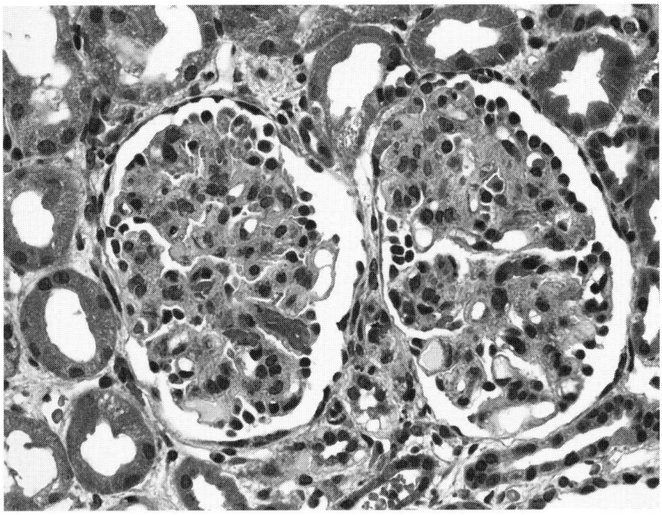

B

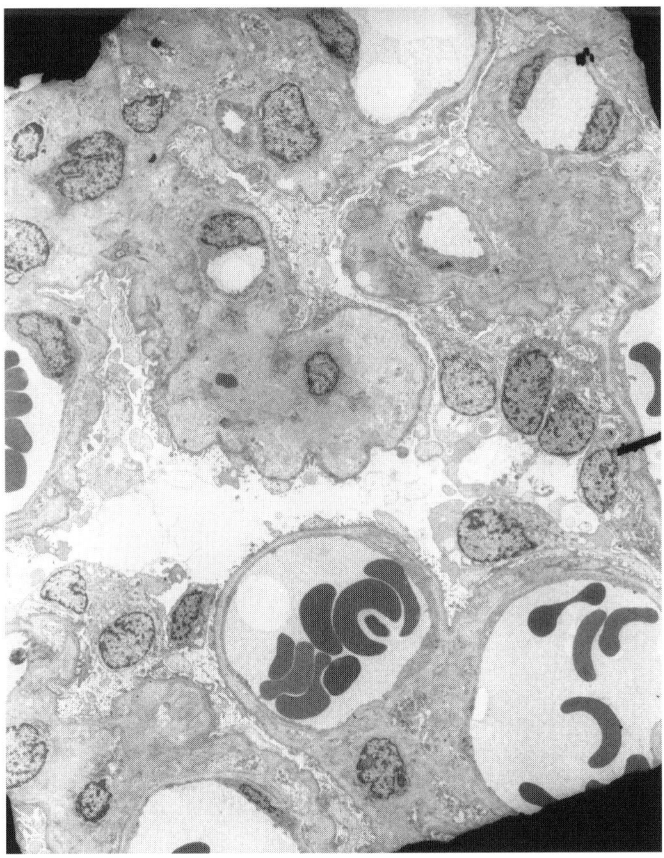

mentemia with low serum levels of the central complement component C3 in patients with familial (atypical) HUS/TTP had been demonstrated [5, 60–62]. In a large study, Noris et al. [62] found an association between (low) C3 serum levels and HUS/TTP within affected families. Furthermore, affected individuals and their healthy relatives showed significantly more Factor H abnormalities than healthy control individuals, and, within families, Factor H abnormalities were significantly correlated with (low) C3 serum levels.

Patients with familial (atypical) HUS showed both autosomal-recessive [63] and autosomal-dominant [1] modes of inheritance. Subsequent studies in patients with familial (atypical) HUS showed mutations in the Factor H gene in some patients, and these mutations were associated with both the spontaneous and the recurrent form of the disease [1, 22, 55, 62–71]. The frequency of Factor H gene mutations was described in patients with sporadic and familial aHUS in a registry of German speaking countries: 14% of 111 patients had Factor H gene mutations including 2 of 8 patients with familial aHUS [49]. In contrast to these data, Wyatt et al. [72] report an allelic frequency of factor H deficiency of 1:326 healthy Caucasian controls from Kentucky, allowing the calculation of the prevalence of homozygous factor H deficiency as 1:425,104 in the United States [9, 72]. The vast majority of the mutations found in Factor H-associated HUS is heterozygous, resulting in one normal and one defective Factor H allele. Factor H plasma levels of these patients can be normal, but also reduced, or even increased [73]. Only very few patients with Factor H-associated HUS carry (genetically or functionally) homozygous mutations with usually no or very little plasma Factor H [65, 66, 68, 69].

Recent progress in the understanding of Factor H protein function revealed the role of Factor H mutations for endothelial cell damage, which is the central feature of the pathogenesis of aHUS. The secreted Factor H protein is composed of 20 repetitive domains termed short consensus repeats (SCR) or complement control proteins (CCP). Structure-function analysis demonstrated that N-terminal SCRs 1–4 are required for complement regulatory activities ("regulatory region") and that C-terminal SCRs 19–20, which are highly conserved within the Factor H protein, play a crucial role for protein function ("recognition region"): they include overlapping heparin and C3b binding domains, and are the binding domain for microbial surface proteins. In addition, Factor H has three C3 binding domains (SCRs 1–4 bind C3b; SCRs 12–14 bind C3c; and SCRs 16–20 bind C3b/C3d) and three heparin binding domains (SCRs 7, 13 and 20) [35, 73, 74].

Factor H gene mutations identified in Factor H-associated HUS cluster in the C terminus of Factor H protein, mainly SCR 20, identifying this domain as a hot spot, but they also occur in almost all of the other SCRs [34, 49, 65, 66, 68–70, 73, 75, 76].

Reflecting the role of Factor H as a complement regulator and the functional organization of the Factor H protein, the current understanding of the pathogenesis of Factor H-associated HUS can be summarized as follows. Endothelial cells coating the inner surface of blood vessels are in direct contact with plasma and are – under physiological conditions – coated with Factor H. The interaction of Factor H with cell surface structures is mediated via the C terminus, and the initial cell surface contact exposes the complement regulatory N-terminal domain of Factor H protein [37, 76–78]. Thus, in addition to its activity in fluid phase, Factor H contributes to the assembly of complement control systems on cell surfaces consisting of two layers: the inner layer activity on the cell membrane level is mediated by membrane-anchored regulators (e.g., MCP, DAF), the outer layer activity in about 20–140 nm (defined by molecule dimensions) distance from the cell membrane by surface-attached soluble regulators such as Factor H [37].

However, Factor H proteins with mutations within SCR 20 – purified from patient plasma or designed as recombinant mutant proteins – show reduced affinity to the central complement component C3b/C3d, to heparin and to endothelial cells, and are inactive [21, 37, 39]. This is in agreement with recent three-dimensional structure studies of the Factor H protein, which permitted the molecular (functional) interpretation of mutations within SCR 20 [79]. In consequence, local activation of the complement system occurs, resulting in a vicious circle of damage of endothelial cells – exposure of the basement membrane – and continued activation of the complement (and the coagulation) system [34, 37, 59].

Under physiological conditions, the concerted action of membrane-anchored and surface-attached fluid-phase regulators provides sufficient complement control and protection of host cells. During situations of "biological stress", like inflammation or microbial infection, and other triggering events like pregnancy, vaccination, or the action of drugs (e.g., calcineurin inhibitors), complement activation occurs rapidly and heavily, and needs to be restricted. Under these conditions the activity of the endogenous, membrane-anchored regulators must be supplemented by surface-attached, fluid-phase regulators to provide sufficient protection of host cells. If such additional regulators are not available, e.g., Factor H deficiency, unrestricted complement activation and subsequently damage occurs [37, 42, 46]. According to a recent report of Dragon-Durey et al. [27], anti-Factor H autoantibodies can also cause defective Factor H function, and can therefore also be responsible for the development of aHUS.

In summary, either reduced availability or functional impairment of Factor H protein critically increases the vulnerability of blood vessel endothelial cells with respect to an attack of the complement system, finally resulting in aHUS.

By contrast, we recently described a patient who developed aHUS on the background of a novel homozygous Factor H mutation in SCR 15. This mutation allows intracellular expression of a truncated Factor H protein, which cannot be secreted, resulting in very low Factor H plasma levels of < 1 µg/mL (normal: about 500 µg/mL). This case is of special interest since despite missing plasma Factor H the patient developed aHUS and not MPGN II, as one would have expected from current experiences and knowledge [68].

However, not only Factor H deficiency but also deficiency in MCP (CD46) causes aHUS, as has recently been shown in patients with a family history of aHUS on the background of complement activation but normal Factor H [6–8]. Like soluble complement regulator Factor H, transmembrane complement regulator MCP inhibits complement activation by inactivating the C3b, which is deposited on target surfaces [6]. Different MCP gene mutations (deletion resulting in a premature stop codon and loss of the C terminus [7]; deletion resulting in intracellular retention of the mutant protein; substitution resulting in the expression of a functionally impaired protein [8]) result in reduced protein expression or impaired protein function were found. These MCP mutations have the same functional consequences as Factor H deficiency: unrestricted activation of the alternative complement pathway [6] (see also chapter by Goodship et al.).

Appendix

Besides deficiency of factors regulating the activation of the alternative complement pathway, e.g., complement Factor H and MCP, there is a second known cause for the development of aHUS: deficiency of von Willebrand factor-cleaving protease (vWF-CP), a factor of the coagulation system [1, 7, 46, 53, 63]. VWF-CP is a vWF-specific metalloprotease of the *ADAMTS* (a disintegrin and metalloprotease, with thrombospondin-1-like domains) family, which physiologically cleaves vWF after synthesis and secretion by endothelial cells. If vWF-CP is deficient, highly reactive ultra-large vWF multimers (ULvWFM) persist, and mediate adhesion and aggregation of circulating platelets mainly at sites of high intravascular shear stress. As a result, large, potentially occlusive platelet-fibrin-rich thrombi appear, which are carried with the bloodstream to ultimately occlude the peripheral microvasculature, i.e., arterioles and capillaries [46, 80, 81].

A severe decrease in activity below detection level (normal: 50–150%) of vWF-CP (*ADAMTS13*) has been described in patients with TTP [81–83]. VWF-CP deficiency can be caused by IgG autoantibodies against vWF-CP, resulting in the (sporadically occurring) acquired form [84–86], or by homozygous or compound heterozygous mutations of the *ADAMTS13* gene, resulting in the (recessively) inherited form [82, 83, 87, 88].

Main target of TTP is the central nervous system with neurological symptoms caused by cerebral ischemia, but different organ systems like the gastrointestinal tract and the kidneys can also be affected. In this case the disease is called TTP. TTP is a severe disease with a mortality rate of > 90% without treatment. When treated with plasma infusion or plasma exchange, however, 60–90% of the patients survive the acute episodes. Relapses occur in about 30–35% of the patients who achieve remission, and a subset of patients develops chronic TTP requiring long-term plasma exchange. If, however, the kidneys are the main target, with renal failure as predominant clinical feature, the disease is called aHUS [46].

Clinical distinction between TTP and aHUS is not always clear-cut: renal abnormalities in patients with a presumed diagnosis TTP, and extrarenal manifestations in patients with a presumed diagnosis aHUS do occur [46]. Some authors, therefore, suggest the term TTP/aHUS for this group of diseases [42, 46]. By contrast, others reject this summarizing term, since assay of *ADAMTS13* activity distinguishes TTP from aHUS and other types of TMA [53]. With respect to future pathophysiology-based classifications, it is important that Noris et al. recently described TTP as a result of two heterozygous *ADAMTS13* gene mutations in two siblings, and, in addition, chronic renal failure in one sibling, possibly caused by a heterozygous Factor H gene mutation [89]. This finding might open the door for an understanding of the pathophysiology of aHUS in which different concepts of explanation – vWF-CP deficiency and RCA deficiency – possibly merge to the concept of one fundamental pathomechanism of endothelial cell damage.

Membranoproliferative glomerulonephritis type II

Even before a link with aHUS was appreciated, dysregulation of the (alternative pathway of the) complement system had been associated with the development of MPGN II [54]. MPGN II, also termed "dense-deposit disease", is a rare disease, which is characterized by complement containing dense deposits within the basement membrane of the glomerular capillary wall, and followed by capillary wall thickening, mesangial cell proliferation and glomerular fibrosis (Fig. 4) [9, 90].

Besides MPGN II, there are two other MPGN subtypes, MPGN I and MPGN III. All three subtypes are characterized by mesangial cell proliferation and increase in mesangial matrix combined with a thickening of the glomerular capillary walls (MPGN I: interposition of mesangial cells and matrix between basement membrane and endothelial cells, resulting in the formation of a double layer structure of subendothelial electron-dense deposits; MPGN III: subendothelial and subepithelial electron-dense deposits). Deposits in all subtypes contain C3 and other complement factors [90]. In some patients combination of MPGN with extrarenal manifestations like lipodystrophy and retina alterations can be found [90, 91].

MPGN mainly affects older children and adolescents (median age at onset of disease: about 10 years). Of the patients, 50% present with nephrotic syndrome, the others with mild proteinuria, 20% with macrohematuria; 30% of the patients develop hypertension at onset of disease. Children with MPGN have an unfavorable late prognosis, and develop ESRD after about 8–16 years (MPGN I: 15.3 years; MPGN II: 8.7 years; MPGN III: 15.9 years) [90, 92].

The development of MPGN II has been linked to the dysregulation of the (alternative pathway of the) complement system caused by the following five different

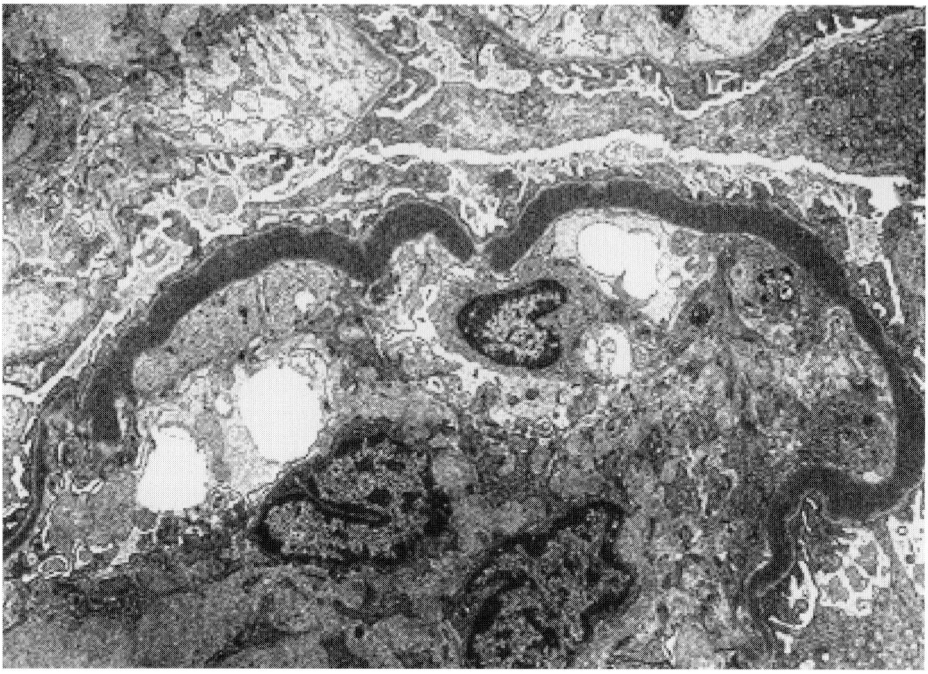

Figure 4
MPGN II
Electron micrograph with segmental osmiophilic transformation of the glomerular basement membrane ("dense deposits") (EM by Rüdiger Waldherr, Heidelberg, Germany).

conditions [9]: (1) congenital absence of Factor H in plasma [5, 10, 13, 66, 72, 93–95]; (2) functional defect of Factor H protein caused by a mutation, e.g., in the regulatory domain [26]; (3) inactivation of Factor H caused by a circulating inhibitor [96]; (4) presence of the IgG autoantibody C3 nephritic Factor (C3NeF) [40, 41, 97, 98]; and (5) mutation in C3 altering its Factor H binding site [99].

Absence of Factor H in plasma as the cause of MPGN II has been observed in humans [5, 10, 13, 66, 72, 93–95], in naturally mutant Factor H-deficient pigs [100, 101], and in genetically engineered Factor H-knockout mice [102]. All mutations so far known result in a block of protein secretion. The mutant protein is expressed in hepatocytes, but is retained in the endoplasmic reticulum, thereby accumulating in the cytoplasm. The absence of Factor H in plasma causes sustained activation of the alternative complement pathway, reflected by consumption of C3 and accumulation of the C3 degradation product C3d in plasma [103].

We recently described siblings with MPGN II who showed activation of the alternative pathway of the complement system despite normal plasma Factor H levels. Genetic analyses revealed a novel mutation of the Factor H gene (cDNA 743–745) causing deletion of Lys in position 224 (ΔK_{224}) located within SCR 4 in both patients. Testing of mutant Factor H protein function revealed normal C-terminal (i.e., binding to heparin, C3d, and endothelial cells), but impaired N-terminal (i.e., cofactor and decay-accelerating activity, binding to central complement C3b) activities [26]. Thus, with respect to complement activation and development of MPGN II functional defect of Factor H protein proves to be of equal importance compared to plasma Factor H deficiency.

Meri et al. [96] isolated a factor from serum and urine of a patient with hypocomplementemic MPGN II (atypical histology with additional subendothelial deposits: intermediate type between MPGN I and II) that was associated with dysfunction of the alternative complement pathway. When mixed with fresh normal serum, the patient's serum induced almost complete conversion of C3. This activity was due to a circulating factor, which was different from C3NeF, but was observed to interact directly with Factor H and therefore can be considered as Factor H antibody.

However, not only the absence (Factor H mutations and Factor H knockout) or inactivation (Factor H antibodies) of Factor H, but also the presence of the IgG autoantibody C3NeF in plasma is linked to MPGN II [40, 41]: 54% of adult and 82% of pediatric MPGN II patients are positive for C3NeF [104]. However, C3NeF is observed not only in patients with MPGN II, but also in patients with the MPGN subtypes I and III*, partial lipodystrophy, retina alterations, meningococcal meningitis and even in healthy individuals [91, 104]. Moreover, it is possible that the same patient is C3NeF positive at one time point, and becomes negative at a different time point [104]. Both Factor H and C3NeF control the same enzyme, the alternative pathway C3 convertase C3bBb. However, while Factor H dissociates the C3bBb complex and – together with Factor I – inactivates the convertase, C3NeF displays the opposite effect: C3NeF stabilizes the convertase and increases its half-life approximately tenfold, a defect which also causes unrestricted activation of the (alternative pathway of the) complement system (Fig. 2) [40, 105].

Marder et al. [99] described a mutation in C3 that alters its Factor H binding site, and renders Factor H unable to dissociate the alternative pathway convertase C3bBb. In consequence, this defect, called Marder's disease, also results in unrestricted activation of the (alternative pathway of the) complement system.

In summary, quite distinct abnormalities can cause dysregulation of the activation of the (alternative pathway of the) complement system. A redundant system of soluble and membrane-anchored RCA prevents unrestricted progression of the activation of the complement system. At particular sites, however, tissue surfaces lack

* Distribution of C3NeF in serum depending on different histological types of MPGN: MPGN II > MPGN I > MPGN III [92].

such additional membrane-anchored regulators so that these structures depend on soluble regulators like Factor H. The glomerular basement membrane represents such a sensitive structure because it lacks endogenous, membrane-anchored regulators [37, 100]. Lack or functional inactivation of Factor H results in continued C3 deposition within the "lamina densa" of the glomerular basement membrane, resulting in "dense-deposit disease" (MPGN II) with thickening of the glomerular basement membrane, impairment of the glomerular filter, hematuria and proteinuria, and, finally, progressive loss of renal function and development of ESRD.

Diagnosis and therapy

Diagnosis

Based on the explanation of aHUS and MPGN II as complement based diseases, the detailed evaluation of the complement system in a stepwise approach is necessary whenever the diagnosis aHUS or MPGN II (or of another disease which is linked to the complement system) is suspected or diagnosed, and other causes – especially for aHUS – are excluded.

In a first step, C3, C3d and APH50 and CH50 [tests of the activation status of the alternative (APH50) and of the classical (CH50) pathway of the complement system] are measured. Low C3 and high C3d levels accompanied by low values for APH50 and CH50 indicate the activation of the (alternative and the classical pathway of the) complement system. If the complement system is activated, plasma Factor H, the major known complement regulator, should be examined. If Factor H is not detectable in plasma or if its level is severely decreased, the Factor H gene should be sequenced.

If, however, the concentration of plasma Factor H is normal despite activation of the (alternative pathway of the) complement system, function of Factor H protein should be examined. If function of Factor H protein is normal, it is likely that one of the other complement regulators, e.g., Factor I or MCP, is responsible for complement activation and, thus, possibly responsible for the disease. If, however, function of Factor H protein is defective, sequencing of Factor H gene should be performed to detect possible Factor H gene mutations, which affect proper function of Factor H protein. Figure 5 summarizes this stepwise approach and provides a diagnostic algorithm for complement based renal diseases.

Whenever unrestricted activation of the (alternative pathway of the) complement system caused by deficiency or defect of one or more complement regulators is identified as cause for aHUS or MPGN II, substitution of this missing or functionally defective factor, e.g., via plasma infusion, is one possible therapeutic option independent of the disease caused by the complement defect.

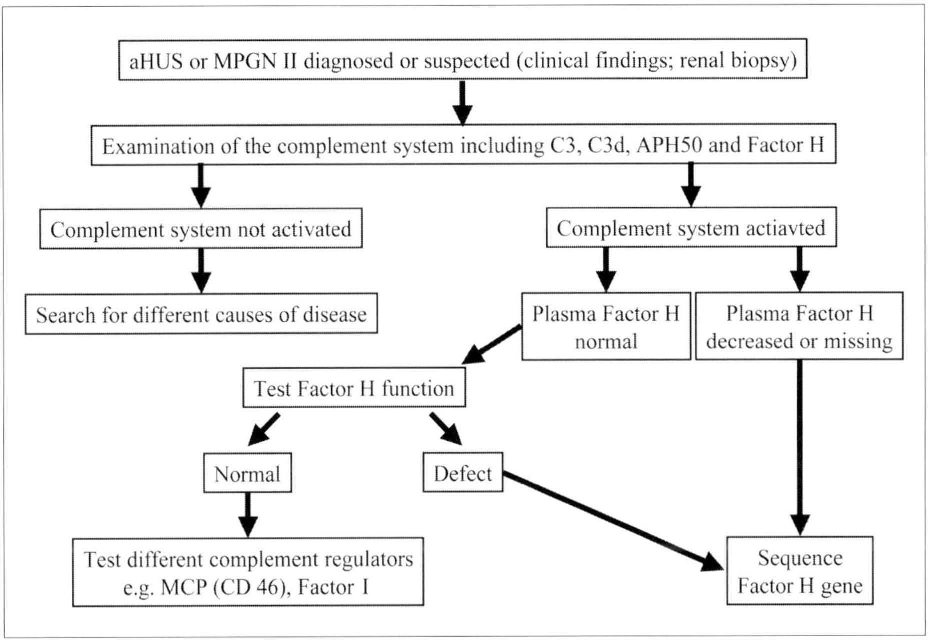

Figure 5
Diagnostic algorithm for complement based renal diseases

Therapy of aUHS

While for classical HUS there is no treatment of proven efficacy besides supportive measures and, if needed, renal replacement therapy, there are several treatment options for complement based, especially Factor H related, aHUS.

Since Factor H is physiologically synthesized by the liver, liver transplantation theoretically represents a causative treatment in patients with Factor H deficiency. In fact, there have been three patients with Factor H-associated aHUS worldwide who underwent (combined kidney and) liver transplantation [106]. However, none of these patients survived and at least one of them developed acute liver failure for severe thrombosis of the liver veins [107].

Restoration of Factor H plasma concentration and/or Factor H function by either plasma exchange or repetitive plasma infusions, however, represents the treatment concept, which is currently accepted [23, 24, 42, 108–110]. However, so far there is rather little secured information about therapy regimens available in the literature, and most authors stress the empirical character of the chosen treatment regimen. Plasma exchange (1–2 plasma volumes/session) or plasma infusion (10–20 to 30–40 mL/kg) is considered first-line therapy. Depending on the need of the patient, plasma infusion is given daily in the beginning, afterwards every 7–14 days. In some patients, daily plasma infusions are required even for a longer period of time [23, 42]. As alternative to fresh frozen plasma (FFP), cryosupernatant – in patients who do not respond to plasma – or solvent detergent-treated plasma – to minimize the risk of viral contamination – can be used [23, 42, 111].

We recently described a patient with complete Factor H deficiency who developed aHUS [68]. Genetic analysis revealed a point mutation of the Factor H gene which introduced a premature stop codon in SCR 15, preventing Factor H protein from being secreted by liver cells. In a time course investigation of this patient following plasma infusion (20 mL/kg), Factor H half-life of about 6 days was determined (Fig. 6A). Based on this result, the patient was treated with regular plasma infusions in cycles of 10–14 days, which is used in the majority of patients reported in the literature. Following plasma infusion, plasma Factor H and C3 levels increased, while C3d levels decreased (Fig. 6A), and in parallel APH50 and CH50 levels increased (Fig. 6B). In summary, these findings reflect successful inhibition of the unrestricted activation of the complement system. The long-term observation of indicators for disease activity and renal function showed that both lactate dehydrogenase (LDH) and creatinine (Crea) in serum decreased upon periodical plasma infusion of 20 mL/kg every 14 days, indicating that this treatment was able to stop disease activity and progression (Fig. 7). The patient recovered completely and kidney function remained normal [68]. Shortly after initial aHUS manifestation and initiation of regular treatment with plasma infusion, a minor aHUS relapse with complement activation, signs of hemolysis, and decrease of renal function occurred after pneumococcal vaccination but could be controlled by immediate plasma infusion.

This observation is in accordance with the literature: conditions of "biological stress", e.g., vaccinations and infections, are reported to be triggering events for new aHUS crises [46, 65]. Thus, in addition to the long-term regimen of regular plasma infusions, the patient was prophylactically treated with additional plasma infusions whenever he developed severe airway or gastrointestinal infections. The typical combination of low C3, high C3d, and low APH50 levels indicated activation of the complement system in each of these situations. Plasma infusions (20 mL/kg/day) were then given daily for the period with fever $>38.5°C$. By this regimen unrestricted activation of the complement system could be stopped (Fig. 8A), and development of new aHUS crises could be prevented, which, in the long run, preserved renal function (Fig. 8B) [68].

As rescue therapy for extremely sick patients with frequent relapses or requirement of large amounts of plasma, splenectomy might be considered [42, 112].

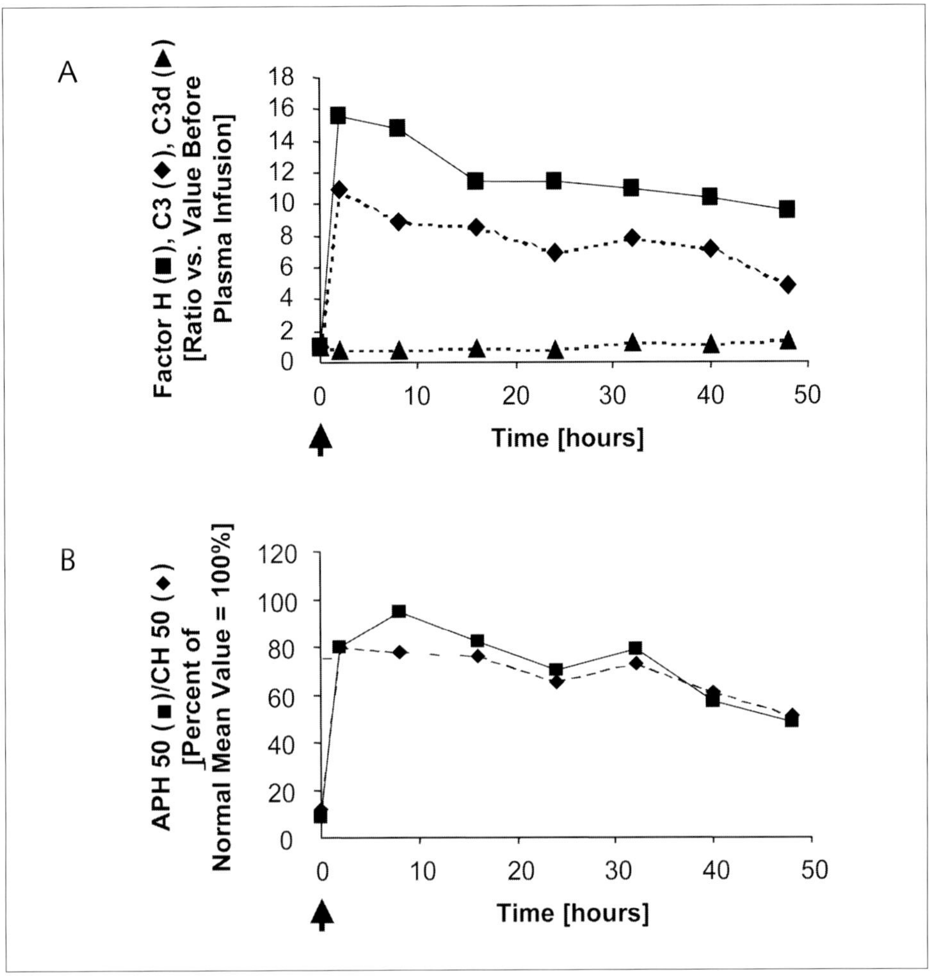

Figure 6
Factor H and complement profile before and following plasma infusion [68]
(A) Factor H, C3 and C3d expressed as ratio of value before plasma infusion. Factor H, C3 and C3d were measured before, and 2, 8, 16, 24, 32, 40 and 48 h after plasma infusion (arrow) by ELISA. The decrease of Factor H in plasma was paralleled by a decrease of C3 and an increase of the C3 degradation product C3d. The patient does not secrete Factor H. Therefore, this decline can be extrapolated to estimate half-life of Factor H in plasma to about 6 days. (B) Complement profile of both the alternative pathway (APH50) and the classical pathway (CH50) expressed as % of normal mean value (=100%) following plasma infusion. Both APH50 and CH50 increased after plasma infusion and decreased gradually afterwards. This indicates complement activation before plasma infusion and proves inhibition of complement activation through plasma infusion (arrow).

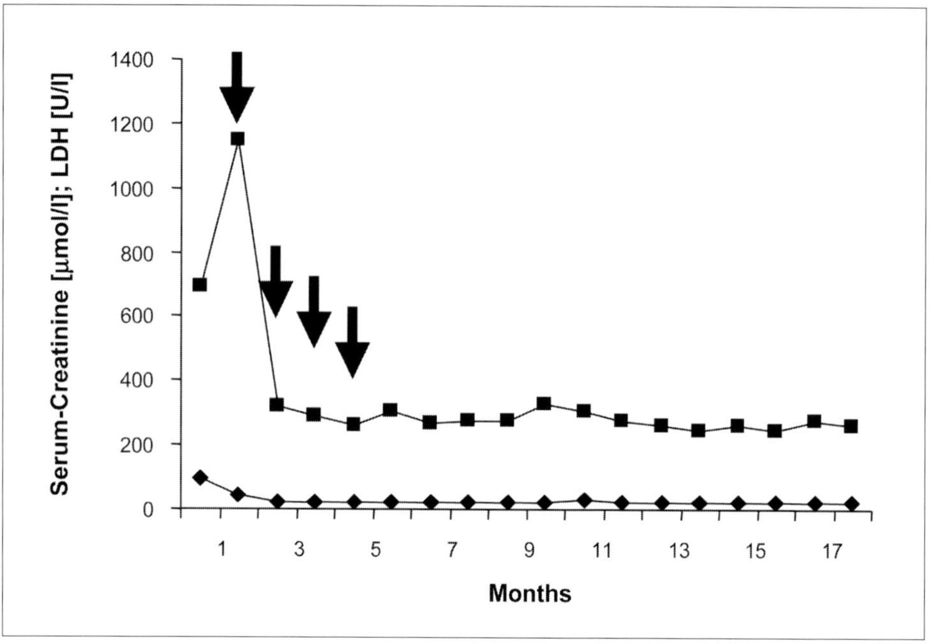

Figure 7
Time course of serum creatinine (diamond) and lactate dehydrogenase (LDH; square)
Upon periodical infusion of FFP (20 mL/kg) every 14 days over a period of 18 months both serum creatinine and LDH decreased immediately and remained within normal range afterwards.

Therapy of MPGN II

A general recommendation for the treatment of MPGN does not exist. Individual treatment concepts include steroids (prednisone), alkylating substances (cyclophosphamide) or other immunosuppressive agents (e.g., mycophenolate mofetil), anti-inflammatory agents (dipyridamole, acetylsalicylic acid, dicumarol) – as monotherapy or in various combinations – and treatment of possible concomitant diseases like hypertension [diuretics, other anti-hypertensive agents, e.g., angiotensin converting enzyme (ACE) inhibitors] [92, 113, 114]. ACE and/or angiotensin (AT) receptor inhibitors can be used to reduce proteinuria and to protect the kidney against the development of tubulointerstitial fibrosis [115, 116].

With the concept of complement-based pathogenesis of MPGN II presented here, however, replacement of those complement factors that are missing or functionally defective, thereby, leading to unrestricted complement activation, becomes a therapeutic option [117, 118]. Thus, periodical plasma infusions, e.g., 10–20 mL/kg

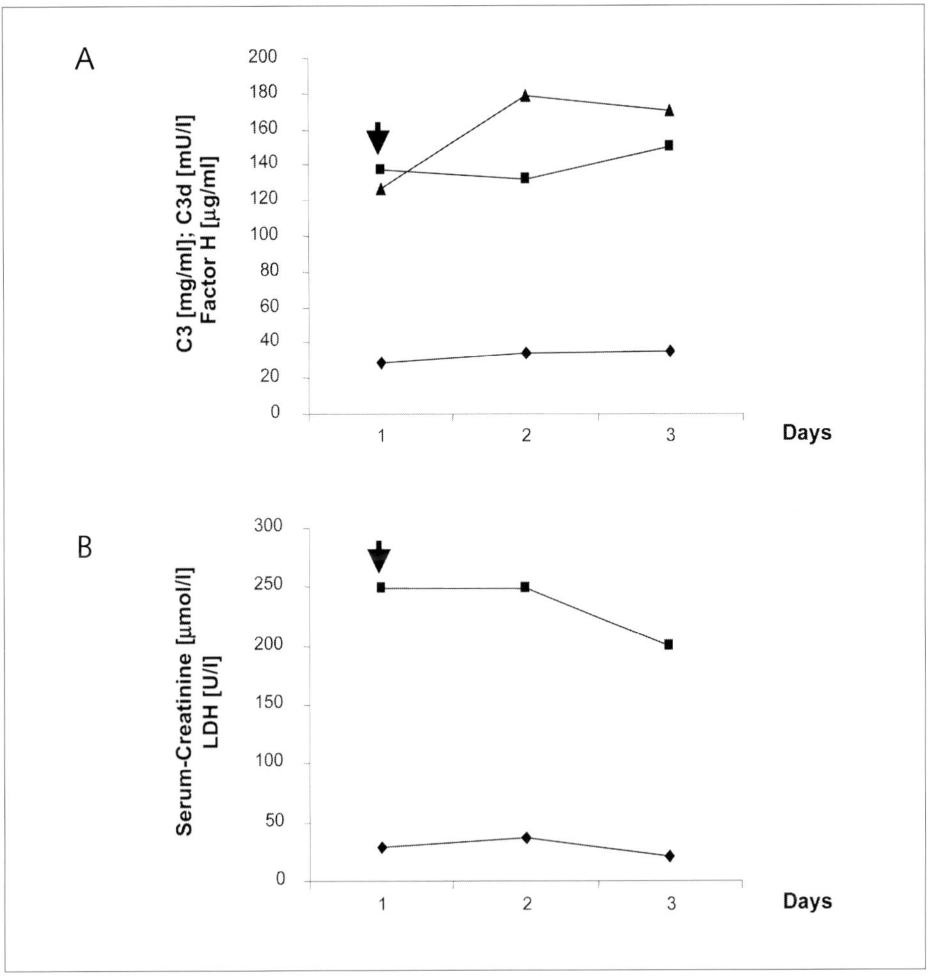

Figure 8
Plasma infusion decreases complement activation in a patient with aHUS on the background of Factor H deficiency triggered by an airway infection
Plasma infusion (arrow) results in (A) an increase of Factor H (triangle) and C3 (diamond), and a decrease of C3d (square), reflecting limitation of the activation of the complement system, and (B) a decrease of both serum creatinine (diamond) and LDH (square), reflecting recovery of renal function.

every 14 days similar to the treatment of aHUS, seem an adequate regimen [26]. This approach is supported by the observation of the beneficial effect of the infusion of porcine plasma to piglets with MPGN II caused by Factor H deficiency [100,

103], and by recently published results obtained in a mouse model of chronic serum sickness, indicating a disease limiting effect of Factor H substitution [119].

In accordance with this concept, we recently published the treatment of two siblings with MPGN II with mild hematuria and proteinuria caused by functional defect of Factor H protein. In these patients complement activation (low C3, high C3d, low APH50 levels) was present despite normal plasma Factor H levels. Patients were treated with regular plasma infusions (20 mL/kg/14 days) to substitute the patient with functionally intact Factor H. Figure 9 demonstrates the immediate effect of this treatment on two serum parameters (C3 and LDH), which reflect the degree of complement activation (Fig. 9A and B) [26]. Acutely, plasma infusion reduced complement activation. Chronically, however, plasma infusion prevented disease progression with development of ESRD.

Similar to aHUS on the background of Factor H deficiency, conditions of "biological stress" like airway or gastrointestinal infections caused complement activation, resulting in a more severe hematuria and proteinuria. Complement activation could successfully be stopped by additional plasma infusions, which were given for the period of fever >38.5°C (Fig. 9a and b).

Besides replacement of deficient or defective complement regulatory factors, plasmapheresis might provide a therapeutic option for cases in which MPGN II is caused by the presence of C3NeF. In support of this approach, Kurtz et al. [120] reported a child with rapidly progressive recurrent MPGN II due to of the presence of C3NeF. In this patient disease progression towards onset of chronic renal failure was delayed by periodical plasmapheresis.

Summary

We here describe aHUS and MPGN II as complement based renal diseases. The increasing number of published case reports now allows a better and more detailed insight into the pathomechanism of both diseases. The general concept consists in a congenital or acquired defect in the regulation of the activation of the alternative complement pathway, leading to self attack and damage of specific structures, i.e., tissue surfaces, endothelial cells of the vasculature, and basement membranes (especially glomerular basement membrane). Besides the known defects in Factor H, MCP, and Factor I, there are more candidate proteins which play a role in complement regulation, e.g., C4BP, Factor B, CR1, DAF, etc., which might in the near future also become identified as factors causing unrestricted alternative complement pathway activation when defective.

In aHUS, complement activation causes lesions of the endothelial cell lineage especially of renal vessels, resulting in thrombembolic occlusion of small vessels followed by the loss of organ function. In MPGN II, however, complement activation causes deposition of complement factors in the "lamina densa" of the glomerular

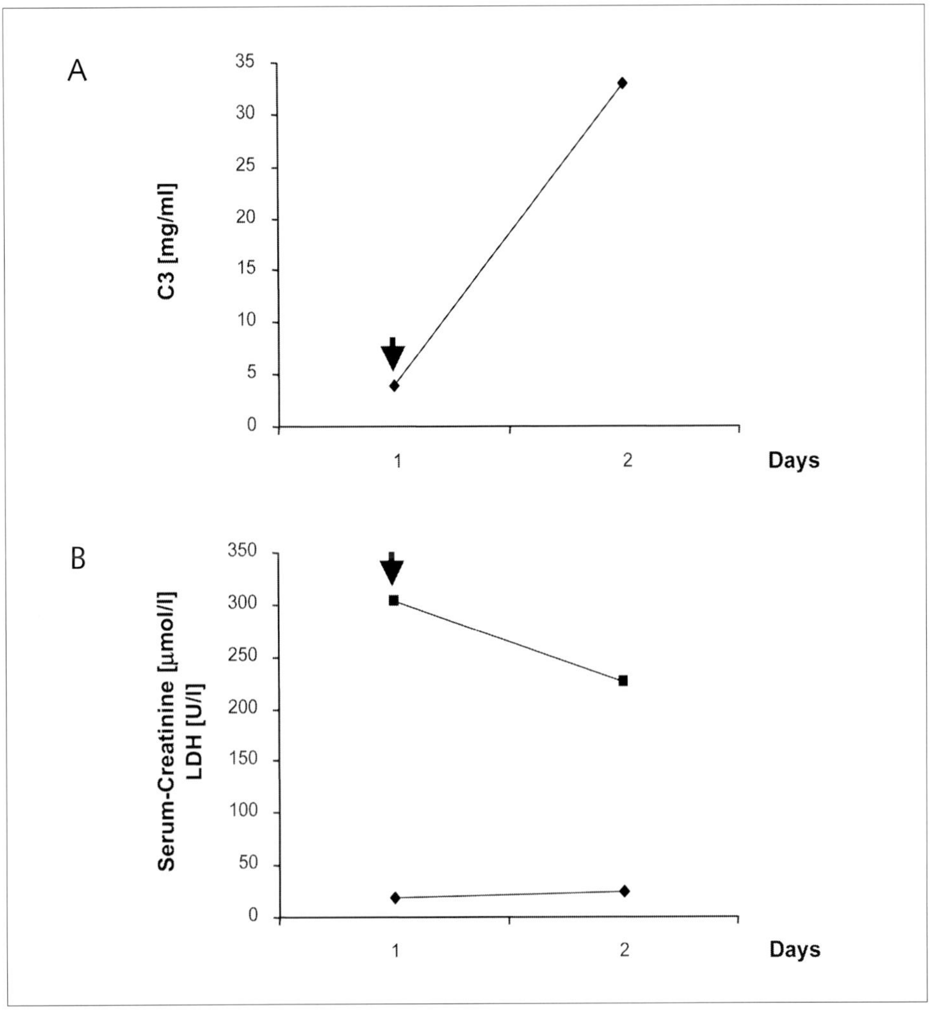

Figure 9
Plasma infusion (arrow) decreases complement activation in a patient with MPGN II on the background of functional Factor H defect
Plasma infusion limits complement activation reflected by an immediate increase in (A) C3 (diamond) and (B) LDH (square). Serum creatinine (diamond) of this patient was normal and was not affected by plasma infusion.

basement membrane (dense-deposit disease), which results in thickening of the basement membrane and impairs the function of the glomerular filter.

While the individual chains of events explaining the pathogenesis of complement-based aHUS and MPGN II are consistent, it is still unclear why different patients with the same defect, e.g., Factor H deficiency, do not develop the same disease but, as in most cases, either aHUS or MPGN II. The two pathologies might be linked by shared pathogenetic mechanisms, or one might be able to evolve into the other [54]. In support of this hypothesis, Mathieson et al. [54] described a family with a father and his son who both developed aHUS in infancy. Both patients showed sustained alternative complement pathway activation. Renal biopsy of the father during adult life showed MPGN (subtype not specified) but it remained open whether the son's course of disease would follow the same pattern.

The plasma levels of alternative complement pathway regulator Factor H might also be of importance. Based on the literature, patients with heterozygous Factor H mutations show reduced plasma levels of Factor H and are prone to develop aHUS, while patients with complete plasma Factor H deficiency on the background of a homozygous Factor H mutation, or patients with loss in protein function, usually develop MPGN II. By contrast, Dragon-Durey et al. [66] identified overlaps between these two groups, describing patients carrying homozygous mutations with no detectable plasma Factor H who develop aHUS and not MPGN II. This observation was confirmed by our own results: we also described a patient with complete Factor H deficiency on the background of a homozygous Factor H mutation who developed aHUS and not MPGN II [68].

Furthermore, not only Factor H deficiency but also functional defects of Factor H cause complement activation, and finally result in disease. Accordingly, we described two siblings with MPGN II with complement activation despite normal levels of plasma Factor H. Genetic analysis identified a novel Factor H gene mutation in the regulatory domain SCR 4 of Factor H, and functional analysis of mutant Factor H protein revealed impaired N-terminal function ("regulation"), while C-terminal function ("activity") was normal [26].

Finally, the variability in the manifestation of these different diseases might also be explainable by the existence of "cofactors", e.g., additional proteins or certain biological stress constellations with catalytic function for disease manifestation.

Substitution of the missing or defective factors (e.g., Factor H) by means of acute or chronic (periodical) plasma infusion represents a therapeutic option for this group of patients with the potential to gain successful recovery from acute renal failure, to maintain normal renal function, and to prevent development of ESRD with the necessity of renal replacement therapy and transplantation.

Regular plasma infusions were, except few mild allergic reactions, usually well tolerated. Hypervolemia and protein overload cause problems only in smaller patients or in situations which require more frequent plasma infusions or the infusion of higher plasma volumes than usual (20 mL/kg) [23].

However, drugs containing pure Factor H protein isolated from plasma or recombinantly synthesized might become a therapeutic instrument for the future.

The use of these drugs would help to reduce the risk of infections and would make treatment of patients more convenient by offering the possibility of subcutaneous or intramuscular application as compared to regular visits to the outpatient clinic for intravenous plasma infusions as currently required.

References

1 Warwicker P, Goodship THJ, Donne RL, Pirson Y, Nicholls A, Ward RM, Turnpenny P, Goodship JA (1998) Genetic studies into inherited and sporadic hemolytic uremic syndrome. *Kidney Int* 53: 836–844

2 Ohali M, Shalev H, Schlesinger M, Katz Y, Kachko L, Carmi R, Sofer S, Landau D (1998) Hypocomplementemic autosomal recessive hemolytic uremic syndrome with decreased factor H. *Pediatr Nephrol* 12: 619–624

3 Pichette V, Quérin S, Schürch W, Brun G, Lehner-Netsch G, Delâge JM (1994) Familial hemolytic-uremic syndrome and homozygous factor H deficiency. *Am J Kidney Dis* 24: 936–941

4 Roodhooft AM, McLean RH, Elst E, Van Acker KJ (1990) Recurrent haemolytic uraemic syndrome and acquired hypomorphic variant of the third component of complement. *Pediatr Nephrol* 4: 597–599

5 Thompson RA, Winterborn MH (1981) Hypocomplementaemia due to a genetic deficiency of beta 1H globulin. *Clin Exp Immunol* 46:110–119

6 Goodship TH, Liszewski MK, Kemp EJ, Richards A, Atkinson JP (2004) Mutations in CD46, a complement regulatory protein, predispose to atypical HUS. *Trends Mol Med* 10: 226–231

7 Noris M, Brioschi S, Caprioli J, Todeschini M, Bresin E, Porrati F, Gamba S, Remuzzi G (2003) Familial haemolytic uraemic syndrome and an MCP mutation. *Lancet* 362: 1542–1547

8 Richards A, Kemp EJ, Liszewski MK, Goodship JA, Lampe AK, Decorte R, Muslumanoglu MH, Kavukcu S, Filler G, Pirson Y et al (2003) Mutations in human complement regulator, membrane cofactor protein (CD46), predispose to development of familial hemolytic uremic syndrome. *Proc Natl Acad Sci USA* 100: 12966–12971

9 Ault BH (2000) Factor H and the pathogenesis of renal disease. *Pediatr Nephrol* 14: 1045–1053

10 Brai M, Misiano G, Maringhini S, Cutaja I, Hauptmann G (1988) Combined homozygous factor H and heterozygous C2 deficiency in an Italian family. *J Clin Immunol* 8: 50–56

11 Fijen C, Kuijper E, Holdrinet A, Daha M, Dankert J (1992) Factor H deficiency in a Dutch family. *Immunobiology* 184: 427

12 Lopez-Larrea C, Dieguez M, Enguix A, Dominguez O, Marin B, Gomez F (1987) A familial deficiency of complement factor H. *Biochem Soc Trans* 15: 648–649

13 Nielsen HE, Christensen KC, Koch C, Thomas BS, Heegard NH, Tranum-Jensen J (1989) Hereditary, complete deficiency of complement factor H associated with recurrent meningococcal disease. *Scand J Immunol* 30: 711–718

14 Rougier N, Kazatchkine MD, Rougier JP, Fremeaux-Bacchi V, Blouin J, Deschenes G, Soto B, Baudouin V, Pautard B, Proesmans W et al (1998) Human complement factor H deficiency associated with hemolytic uremic syndrome. *J Am Soc Nephrol* 9: 2318–2326

15 Friese MA, Hellwage J, Jokiranta TS, Meri S, Müller-Quernheim HJ, Peter HH, Eibel H, Zipfel PF (2000) Different regulation of factor H and FHL-1/reconectin by inflammatory mediators and expression of the two proteins in rheumatoid arthritis (RA). *Clin Exp Immunol* 121: 406–415

16 Friese MA, Manuelian T, Junnikkala S, Hellwage J, Meri S, Peter HH, Gordon DL, Eibel H, Zipfel PF (2003) Release of endogenous anti-inflammatory complement regulators FHL-1 and factor H protects synovial fibroblasts during rheumatoid arthritis. *Clin Exp Immunol* 132: 485–495

17 Junnikkala S, Hakulinen J, Jarva H, Manuelian T, Bjorge L, Butzow R, Zipfel PF, Meri S (2002) Secretion of soluble complement inhibitors factor H and factor H-like protein (FHL-1) by ovarian tumor cells. *Br J Cancer* 87: 1119–1127

18 Junnikkala S, Jokiranta TS, Friese MA, Jarva H, Zipfel PF, Meri S (2000) Exceptional resistance of human H2 glioblastoma cells to complement-mediated killing by expression and utilization of factor H and factor H-like protein 1. *J Immunol* 164: 6075–6081

19 Devine DV, Siegel RS, Rosse WF (1987) Interactions of the platelets in paroxysmal nocturnal hemoglobinuria with complement. Relationship to defects in the regulation of complement and to platalet survival *in vivo*. *J Clin Invest* 79: 131–137

20 DiScipio RG, Daffern PJ, Schraufstatter IU, Sriramarao P (1998) Human polymorphonuclear leukocytes adhere to complement factor H through an interaction that involves alphaMbeta2 (CD11b/CD18). *J Immunol* 160: 4057–4066

21 Manuelian T, Hellwage J, Meri S, Caprioli J, Noris M, Heinen S, Jozsi M, Neumann HPG, Remuzzi G, Zipfel PF (2003) Mutations in factor H reduce binding affinity to C3b and heparin and surface attachment to endothelial cells in hemolytic uremic syndrome. *J Clin Invest* 111: 1181–1190

22 Berner R, Krause MF, Gordjani N, Zipfel PF, Boehm N, Krueger M, Brandis M, Zimmerhackl LB (2002) Hemolytic uremic syndrome due to an altered factor H triggered by neonatal pertussis. *Pediatr Nephrol* 17: 190–192

23 Filler G, Radhakrishnan S, Strain L, Hill A, Knoll G, Goodship TH (2004) Challenges in management of infantile factor H associated haemolytic uremic syndrome. *Pediatr Nephrol* 19: 908–911

24 Gerber A, Kirchhoff-Moradpour AH, Obieglo S, Brandis M, Kirschfink M, Zipfel PF, Goodship JA, Zimmerhackl LB (2003) Successful (?) therapy of hemolytic-uremic syndrome with factor H abnormality. *Pediatr Nephrol* 18: 952–955

25 Olie KH, Florquin S, Groothoff JW, Verlaak R, Strain L, Goodship THJ, Weening JJ,

Davin JC (2004) Atypical relapse of hemolytic uremic syndrome after transplantation. *Pediatr Nephrol* 19: 1173–1176

26 Licht C, Heinen S, Józsi M, Löschmann I, Saunders RE, Perkins SJ, Waldherr R, Skerka C, Kirschfink M, Hoppe B, Zipfel PF (2005) Deletion of a single amino acid (K224) in the regulatory domain of factor H reveals a novel pathomechanism for membranoproliferative glomerulonephritis type II. *Submitted*

27 Dragon-Durey MA, Loirat C, Cloarec S, Macher MA, Blouin J, Nivet H, Weiss L, Fridman WH, Frémeaux-Bacchi V (2005) Anti-factor H autoantibodies associated with atypical hemolytic uremic syndrome. *J Am Soc Nephrol* 16: 555–563

28 Kavanagh D, Kemp EJ, Mayland E, Winney RJ, Duffield JS, Warwick G, Richards A, Ward R, Goddship JA, Goddship TH (2005) Mutations in complement factor I predispose to development of atypical hemolytic uremic syndrome. *J Am Soc Nephrol* 16: 2150-2155

29 Abbas AK, Lichtman AH, Pober JS (2000) Effector mechanisms of humoral immunity. In: *Cellular and Molecular Immunology*, 4th edition. WB Saunders Company, Philadelphia, 309–339

30 Walport MJ (2001) Complement (first of two parts). *N Engl J Med* 344: 1058–1066

31 Walport MJ (2001) Complement (second of two parts). *N Engl J Med* 344: 1140–1144

32 Young NS (2005) Paroxysmal nocturnal hemoglobinuria: current issues in pathophysiology and treatment. *Curr Hematol Rep* 4: 103–109

33 Zipfel PF (2001) Complement factor H: physiology and pathophysiology. *Semin Thromb Hemost* 27: 191–199

34 Zipfel PF (2001) Hemolytic uremic syndrome: how do factor H mutants mediate endothelial damage? *Trends Immunol* 22: 345–348

35 Zipfel PF, Jokiranta TS, Hellwage J, Koistinen V, Meri S (1999) The factor H protein family. *Immunopharmacology* 42: 53–60

36 Zipfel PF, Skerka C (1999) FHL-1/reconectin: a human complement and immune regulator with cell-adhesive function. *Immunol Today* 20: 135–140

37 Józsi M, Manuelian T, Heinen S, Oppermann M, Zipfel PF (2004) Attachment of the soluble complement regulator factor H to the cell and tissue surfaces: relevance for pathology. *Histol Histopathol* 19: 251–258

38 Quigg RJ (2003) Complement and the kidney. *J Immunol* 171: 3319–3324

39 Sánchez-Corral P, Pérez-Caballero D, Huarte O, Simckes AM, Goicoechea E, López-Trascasa M, Rodríguez de Córdoba S (2002) Structural and functional characterization of factor H mutations associated with atypical hemolytic uremic syndrome. *Am J Hum Genet* 71: 1285–1295

40 Daha MR, Fearon DT, Austen KF (1976) C3nephritic factor (C3NeF): stabilization of fluid phase and cell-bound alternative pathway convertase. *J Immunol* 116: 1–7

41 Mathieson PW, Würzner R, Oliviera DBG, Lachmann PJ, Peters DK (1993) Complement-mediated adipocyte lysis by nephritic factor sera. *J Exp Med* 177: 1827–1831

42 Ruggenenti P, Noris M, Remuzzi G (2001) Thrombotic microangiopathy, hemolytic uremic syndrome, and thrombotic thrombocytopenic purpura. *Kidney Int* 60: 831–846

43 Symmers WS (1952) Thrombotic microangiopathic haemolytic anemia (thrombotic microangiopathy). *Br Med J* 2: 897–903
44 Gasser C, Gautier E, Steck A, Siebenmann RE, Oechslin R (1955) Hemolytic-uremic syndrome: bilateral necrosis of the renal cortex in acute acquired hemolytic anemia. *Schweiz Med Wochenschr* 85: 905–909
45 Moschcowitz E (1924) Hyaline thrombosis of the terminal arterioles and capillaries: a hitherto undescribed disease. *Proc NY Pathol Soc* 24: 21–24
46 Moake JL (2002) Thrombotic microangiopathies. *N Engl J Med* 347: 589–600
47 Donne RL, Abbs I, Barany P, Elinder CG, Little M, Conlon P, Goodship THJ (2002) Recurrence of hemolytic uremic syndrome after live related renal transplantation associated with subsequent de novo disease in the donor. *Am J Kidney Dis* 40: E22
48 Loirat C, Niaudet P (2003) The risk of recurrence of hemolytic uremic syndrome after renal transplantation in children. *Pediatr Nephrol* 18: 1095–1101
49 Neuman HPH, Salzmann M, Bohnert-Iwan B, Mannuelian T, Skerka C, Lenk D, Bender BU, Cybulla M, Riegler P, Königsrainer A et al (2003) Haemolytic uraemic syndrome and mutations of the factor H gene: a registry-based study of German speaking countries. *J Med Genet* 40: 676–681
50 Taylor CM, Chua C, Howie AJ, Ridson RA (2004) Clinico-pathological findings in diarrhoea negative haemolytic uraemic syndrome. *Pediatr Nephrol* 19: 419–425
51 Hughes AK, Ergonul Z, Stricklett PK, Kohan DE (2002) Molecular basis for high renal cell sensitivity to the cytotoxic effects of shigatoxin-1: upregulation of globotriaosylceramide expression. *J Am Soc Nephrol* 13:2239-2245
52 Hughes AK, Stricklett PK, Schmid D, Kohan DE (2000) Cytotoxic effect of Shiga toxin-1 on human glomerular epithelial cells. *Kidney Int* 57:2350-2359
53 Tsai HM (2003) Advances in the pathogenesis, diagnosis, and treatment of thrombotic thrombocytopenic purpura. *J Am Soc Nephrol* 14: 1072–1081
54 Mathieson P (2002) Complement factor H and haemolytic uraemic syndrome. *Lancet* 359: 801–802
55 Richards A, Goodship JA, Goodship TA (2002) The genetics and pathogenesis of haemolytic uraemic syndrome and thrombotic thrombocytopenic purpura. *Curr Opin Nephrol Hypertens* 11: 431–435
56 Taylor CM (2001) Complement factor H and the haemolytic uraemic syndrome. *Lancet* 358: 1200–1202
57 Taylor CM (2001) Hemolytic-uremic syndrome and complement factor H deficiency: clinical aspects. *Semin Thromb Hemost* 27: 185–190
58 Zipfel PF, Hellwage J, Friese MA, Hegasy G, Jokiranta ST, Meri S (1999) Factor H and disease: a complement regulator affects vital body function. *Mol Immunol* 36: 241–248
59 Zipfel PF, Skerka C, Caprioli J, Manuelian T, Neumann HH, Noris M, Remuzzi G (2001) Complement factor H and hemolytic uremic syndrome. *Int Immunopharmacol* 1: 461–468
60 Kaplan BS, Chesney RW, Drummond KN (1975) Hemolytic uremic syndrome in families. *N Engl J Med* 292: 1090–1093

61 Kaplan BS, Meyers KE, Schulman SL (1998) The pathogenesis of haemolytic uremic syndrome. *J Am Soc Nephrol* 9: 1126–1133

62 Noris M, Ruggenenti P, Perna A, Orisio S, Caprioli J, Skerka C, Vasile B, Zipfel PF, Remuzzi G (1999) Hypocomplementemia discloses genetic predisposition to hemolytic uremic syndrome and thrombotic thrombocytopenic purpura: role of factor H abnormalities. *J Am Soc Nephrol* 10: 281–293

63 Ying L, Katz Y, Schlesinger M, Carmi R, Shalev H, Haider N, Beck G, Sheffield VC, Landau D (1999) Complement factor H gene mutation associated with autosomal recessive atypical haemolytic uremic syndrome. *Am J Hum Genet* 65: 1538–1546

64 Buddles MRH, Donne RL, Richards A, Goodship J, Goodship THJ (2000) Complement factor H gene mutation associated with autosomal recessive atypical hemolytic uremic syndrome. *Am J Hum Genet* 66: 1721–1722

65 Caprioli J, Bettinaglio P, Zipfel PF, Amadei B, Daina E, Gamba S, Skerka C, Marziliano N, Remuzzi G, Noris M (2001) The molecular basis of familial hemolytic uremic syndrome: mutation analysis of factor H gene reveals a hot spot in short consensus repeat 20. *J Am Soc Nephrol* 12: 297–307

66 Dragon-Durey MA, Frémeaux-Bacchi V, Loirat C, Blouin J, Niaudet P, Deschenes G, Coppo P, Fridman WH, Weiss L (2004) Heterozygous and homozygous factor H deficiencies associated with hemolytic uremic syndrome or membranoproliferative glomerulonephritis: report and genetic analysis of 16 cases. *J Am Soc Nephrol* 15: 787–795

67 Landau D, Shalev H, Levy-Finer G, Polonsky A, Segev Y, Katchko L (2001) Familial hemolytic uremic syndrome associated with complement factor H deficiency. *J Pediatr* 138: 412–417

68 Licht C, Weyersberg A, Heinen S, Stapenhorst L, Devenge J, Beck B, Waldherr R, Kirschfink M, Zipfel PF, Hoppe B (2005) Successful plasma therapy for atypical hemolytic uremic syndrome (aHUS) caused by factor H deficiency due to a mutation in the complement cofactor protein (CCP) domain 15. *Am J Kidney Dis* 45: 415–421

69 Pérez-Caballero D, González-Rubio C, Gallardo ME, Vera M, López-Trascasa M, Rodríguez de Córdoba S, Sánchez-Corral P (2001) Clustering of missense mutations in the C-terminal region of factor H in atypical haemolytic uremic syndrome. *Am J Hum Genet* 68: 478–484

70 Richards A, Buddles MR, Donne RL, Kaplan BS, Kirk E, Venning MC, Tielemans CL, Goodship JA, Goodship THJ (2001) Factor H mutations in hemolytic uremic syndrome cluster in exons 18-20, a domain important for host cell recognition. *Am J Hum Genet* 68:485–490

71 Warwicker P, Donne RL, Goodship JA, Goodship THJ, Howie AJ, Kumararatne DS, Thompson RA, Taylor CM (1999) Familial relapsing haemolytic uraemic syndrome and complement factor H deficiency. *Nephrol Dial Transplant* 14: 1229–1233

72 Wyatt RJ, Julian BA, Weinstein A, Rothfield NF, McLean RH (1982) Partial H (beta 1H) deficiency and glomerulonephritis in two families. *J Clin Immunol* 2: 110–117

73 Zipfel PF, Neumann HPH, Jozsi M (2003) Genetic screening in haemolytic uraemic syndrome. *Curr Opin Nephrol Hypertens* 12: 653–657

74 Zipfel PF, Skerka C, Hellwage J, Jokiranta ST, Meri S, Brade V, Kraiczy P, Noris M, Remuzzi G (2002) Factor H family proteins: on complement, microbes and human diseases. *Biochem Soc Trans* 30: 971–978

75 Caprioli J, Castelletti F, Bucchioni S, Bettinaglio P, Bresin E, Pianetti G, Gamba S, Brioschi S, Daina E, Remuzzi G, Noris M (2003) Complement factor H mutations and gene polymorphisms in haemolytic uraemic syndrome: the C-257T, the A2089G and the G2881T polymorphisms are strongly associated with the disease. *Human Molecular Genetics* 12: 3385–3395

76 Pangburn MK (2002) Localization of the host recognition functions of complement factor H at the carboxyl-terminal: implications for hemolytic uremic syndrome. *J Immunol* 169: 4702–2706

77 Hellwage J, Jokiranta TS, Friese MA, Wolk TU, Kampen E, Zipfel PF, Meri S (2002) Complement C3b/C3d and cell surface polyanions are recognized by overlapping binding sites on the most carboxyl-terminal domain of complement factor H. *J Immunol* 169: 6935–6944

78 Sánchez-Corral P, González-Rubio C, Rodríguez de Córdoba S, López-Trascasa M (2004) Functional analysis in serum from atypical hemolytic uremic syndrome patients reveals impaired protection of host cells associated with mutations in factor H. *Mol Immunol* 41: 81–84

79 Perkins SJ, Goodship THJ (2002) Molecular modelling of the C-terminal domains of factor H of human complement: a correlation between haemolytic uraemic syndrome and a predicted heparin binding site. *J Mol Biol* 316: 217–224

80 Furlan M, Laemmle B (2000) Haemolytic-uraemic syndrome and thrombotic thrombocytopenic purpura – new insights into underlying biochemical mechanisms. *Nephrol Dial Transplant* 15: 1112–1114

81 Furlan M, Robles R, Galbusera M, Remuzzi G, Kyrle PA, Brenner B, Krause M, Scharrer I, Aumann V, Mittler U et al (1998) Von Willebrand factor-cleaving protease in thrombotic-thrombocytopenic purpura and the hemolytic-uremic syndrome. *N Engl J Med* 339: 1578–1584

82 Furlan M, Robles R, Solenthaler M, Wassmer M, Sandoz P, Lammle B (1997) Deficient activity of von Willebrand factor-cleaving protease in chronic relapsing thrombotic thrombocytopenic purpura. *Blood* 89: 3097–3103

83 Schneppenheim R, Budde U, Oyen F, Angerhaus D, Aumann V, Drewke E, Hassenpflug W, Häberle J, Kentouche K, Kohne E et al (2003) Von Willebrand factor cleaving protease and ADAMTS13 mutations in childhood TTP. *Blood* 101: 1845–1850

84 Furlan M, Robles R, Solenthaler M, Lammle B (1998) Acquired deficiency of von Willebrand factor-cleaving protease in a patient with thrombotic thrombocytopenic purpura. *Blood* 91: 2839–2846

85 Rieger M, Mannucci PM, Hovinga JA, Herzog A, Gerstenbauer G, Konetschny C, Zimmermann K, Scharrer I, Peyvandi F, Galbusera M et al (2005) ADAMTS13 autoantibodies in patients with thrombotic microangiopathies and other immunomediated diseases. *Blood* May 17 [Epub ahead of print]

86 Tsai HM, Lian ECY (1998) Antibodies to von Willebrand factor-cleaving protease in acute thrombotic thrombocytopenic purpura. *N Engl J Med* 339: 1585–1594

87 Levy GG, Nichols WC, Lian EC, Foroud T, McClintick JN, Mc Gee BM, Yang AY, Siemieniak DR, Stark KR, Gruppo R et al (2001) Mutations in a member of the ADAMTS gene family cause thrombotic thrombocytopenic purpura. *Nature* 413: 488–494

88 Licht C, Stapenhorst L, Simon T, Budde U, Schneppenheim R, Hoppe B (2004) Two novel ADAMTS13 gene mutations in thrombotic thrombocytopenic purpura/hemolytic-uremic syndrome (TTP/HUS). *Kidney Int* 66: 955–958

89 Noris M, Bucchioni S, Galbusera M, Donadelli R, Bresin E, Castelletti F, Caprioli J, Brioschi S, Scheiflinger F, Remuzzi G (2005) Complement factor H mutation in familial thrombotic thrombocytopenic purpura with ADAMTS13 deficiency and renal involvment *J Am Soc Nephrol* 16: 1177–1183

90 Schärer K (2002) Membranoproliferative Glomerulonephritis. In: K Schärer, O Mehls (eds): *Pädiatrische Nephrologie*. Springer-Verlag, Berlin, 230–232

91 Levy Y, George J, Yona E, Shoenfeld Y (1998) Partial lipodystrophy, mesangiocapillary glomerulonephritis, and complement dysregulation. An autoimmune phenomenon. *Immunol Res* 18: 55–60

92 Schwertz R, de Jong R, Gretz N, Kirschfink M, Anders D, Schärer K and Arbeitsgemeinschaft Pädiatrische Nephrologie (1996) Outcome of idiopathic membranoproliferative glomerulonephritis in children. *Acta Paediatr* 85: 308–312

93 Fijen CA, Kuijper EJ, Te Bulte M, van de Heuvel MM, Holdrinet AC, Sim RB, Daha MR, Dankert J (1996) Heterozygous and homozygous factor H deficiency states in a Dutch family. *Clin Exp Immunol* 105: 511–516

94 Levy M, Halbwachs-Mecarelli L, Gubler MC, Kohout G, Bensenouci A, Niaudet P, Hauptmann G, Lesavre P (1986) H deficiency in two brothers with atypical dense intramembranous deposit disease. *Kidney Int* 30: 949–956

95 Vogt BA, Wyatt RJ, Burke BA, Simonton SC, Kashtan CE (1995) Inherited factor H deficiency and collagen type III glomerulopathy. *Pediatr Nephrol* 9: 11–15

96 Meri S, Koistinen V, Miettinen A, Tornroth T, Seppala IJ (1992) Activation of the alternative pathway of complement by monoclonal lambda light chains in membranoproliferative glomerulonephritis. *J Exp Med* 175: 939–950

97 West CD (1994) Nephritic factors predispose to chronic glomerulonephritis. *Am J Kidney Dis* 24: 956–963

98 West CD, McAdams AJ (1999) The alternative pathway C3 convertase and glomerular deposits. *Pediatr Nephrol* 13: 448–453

99 Marder HK, Coleman TH, Forristal J, Beischel L, West CD (1983) An inherited defect in the C3 convertase, C3b,Bb, associated with glomerulonephritis. *Kidney Int* 23: 749–758

100 Hogasen K, Jansen JH, Mollnes TE, Hovdenes J, Harboe M (1995) Hereditary porcine membranoproliferative glomerulonephritis type II is caused by factor H deficiency. *J Clin Invest* 95: 1054–1061

101 Jansen JH, Hogasen K, Grondahl AM (1995) Porcine membranoproliferative glomerulonephritis type II: an autosomal recessive deficiency of factor H. *Veterinary Record* 137: 240–244
102 Pickering MC, Cook HT, Warren J, Bygrave AE, Moss J, Walport MJ, Botto M (2002) Uncontrolled C3 activation causes membranoproliferative glomerulonephritis in mice deficient in complement factor H. *Nat Genet* 31: 424–428
103 Hegasy GA, Manuelian T, Hogasen K, Jansen JH, Zipfel PF (2002) The molecular basis for hereditary porcine membranoproliferative glomerulonephritis type II. *Am J Pathol* 161: 2027–2034
104 Schwertz R, Rother U, Anders D, Gretz N, Schärer K, Kirschfink M (2001) Complement analysis in children with idiopathic membranoproliferative glomerulonephritis: a long-term follow-up. *Pediatr Allergy Immunol* 12: 166–172
105 Spitzer RE, Stitzel AE, Tsokos GC (1990) Human antiidiotypic antibody responses to autoantibody against alternative pathway C3 convertase. *Clin Immunol Immunopathol* 57: 19–32
106 Remuzzi G, Ruggenenti P, Codazzi D, Noris M, Caprioli J, Locatelli G, Gridelli B (2002) Combined kidney and liver transplantation for familial haemolytic uraemic syndrome. *Lancet* 359: 1671–1672
107 Remuzzi G, Ruggenenti P, Colledan M, Gridelli B, Bertani A, Bettinaglio P, Bucchioni S, Sonzogni A, Bonanomi E, Sonzogni V et al (2005) Hemolytic uremic syndrome: a fatal outcome after kidney and liver transplantation performed to correct factor H gene mutation. *Am J Transplant* 5: 1146–1150
108 Nathanson S, Frémeaux-Bacchi V, Deschênes G (2001) Successful plasma therapy in hemolytic uremic syndrome with factor H deficiency. *Pediatr Nephrol* 16: 554–556
109 Özcakar ZB, Yalcinkaya F, Derelli E, Acar B, Yüksel S, Tulunay Ö, Ekim M (2004) Favorable outcome of haemolytic uremic syndrome with factor H deficiency. *Pediatr Nephrol* 19: 815–816
110 Stratton JD, Warwicker P (2002) Successful treatment of factor H-related haemolytic-uraemic syndrome. *Nephrol Dial Transplant* 17: 684–685
111 George JN (2000) How I treat patients with thrombotic thrombocytopenic purpura-hemolytic uremic syndrome. *Blood* 96: 1223–1229
112 Hayward CP, Sutton DM, Carter WH, Campbell ED, Scott JG, Francombe WH, Shumak KH, Baker MA (1994) Treatment outcomes in patients with adult thrombotic thrombocytopenic purpura-hemolytic uremic syndrome. *Arch Intern Med* 154: 982–987
113 Jones G, Juszczak M, Kingdon E, Harber M, Sweny P, Burns A (2004) Treatment of idiopathic membranoproliferative glomerulonephritis with mycophenolate mofetil and steroids. *Nephrol Dial Transplant* 19: 3160–3164
114 Levin A (1999) Management of membranoproliferative glomerulonephritis: evidence-based recommendations. *Kidney Int* 55: S41–S46
115 Gross O, Beirowski B, Koepke ML, Kuck J, Reiner M, Addicks K, Smyth N, Schulze-Lohoff E, Weber M (2003) Preemptive ramipril therapy delays renal failure and reduces

renal fibrosis in COL4A3-knockout mice with Alport syndrome. *Kidney Int* 63: 438–446

116 Gross O, Schulze-Lohoff E, Koepke ML, Beirowski B, Addicks K, Bloch W, Smyth N, Weber M (2004) Antifibrotic, nephroprotective potential of ACE inhibitor vs AT1 antagonist in a murine model of renal fibrosis. *Nephrol Dial Transplant* 19: 1716–1723

117 Appel GB, Cook HT, Hageman G, Jennette JC, Kashgarian M, Kirschfink M, Lambris JD, Lanning L, Lutz HU, Meri S et al (2005) Membranoproliferative glomerulonephritis type II (dense deposit disease): an update. *J Am Soc Nephrol* 16: 1392–1403

118 Cunningham PN, Quigg RJ (2005) Contrasting roles of complement activation and its regulation in membranous nephropathy. *J Am Soc Nephrol* 16: 1214–1222

119 Alexander JJ, Pickering MC, Haas M, Osawe I, Quigg RJ (2005) Complement factor H limits immune complex deposition and prevents inflammation and scarring in glomeruli of mice with chronic serum sickness. *J Am Soc Nephrol* 16: 52–57

120 Kurtz KA, Schlueter AJ (2002) Management of membranoproliferative glomerulonephritis type II with plasmapheresis. *J Clin Apheresis* 17: 135–137

The role of complement in membranoproliferative glomerulonephritis

Peter F. Zipfel[1], Richard J.H. Smith[2] and Stefan Heinen[1]

[1]Department of Infection Biology, Leibniz Institute for Natural Products Research and Infection Biology, Hans Knoell Institute, Beutenbergstr. 11a, 07745 Jena, Germany;
[2]Department of Otolaryngology, 200 Hawkins Drive, 21151 PFP, The University of Iowa, Iowa City, IA 52242, USA

Introduction

Membranoproliferative glomerulonephritis (MPGN) is a rare disease, and is characterized by complement-containing dense deposits within the basement membrane of the glomerular capillary wall. These changes are followed by capillary wall thickening, mesangial cell proliferation and glomerular fibrosis. This non-immune complex-mediated form of glomerulonephritidis is caused by defective complement control, and is associated with hypocomplementemia due to continuous uncontrolled complement activation. Hypercatabolism of C3 is caused by quite distinct scenarios including: (i) the absence of individual complement components, (ii) an increased turnover of the cascade, or (iii) defective complement control, particularly caused by the absence or functional inactivation of the fluid-phase regulator Factor H or the presence of C3 nephritic factor. Here we discuss the role of complement, particularly of the alternative pathway of complement, in the pathomechanisms of MPGN. We describe and summarize how and why apparently distinct scenarios that affect the activity of the central alternative pathway convertase cause damage to the glomerular basement membrane (GBM) of the kidney.

Membranoproliferative glomerulonephritis: the disease

Complement defects result in kidney diseases, and dysregulation of the alternative pathway, particularly at the level of the amplification convertase C3bBb, causes MPGN. MPGN is a form of glomerulonephritis that is associated with changes of the GBM and mesangial cell proliferation [1, 2]. The disease accounts for about 5% of all cases of idiopathic glomerulopathies, and is caused by an abnormal immune response, with complement and/or immune deposits of antibodies formed in the kidney [3]. MPGN is most common in people under the age of 30 years and both sexes

Complement and Kidney Disease, edited by Peter F. Zipfel

are affected. Complement and immune deposits are formed within the capillaries of the kidney, and cause thickening of the capillaries and impairment of kidney filtration. There is a gradual decline in renal function with 50% of patients reaching end-stage renal failure within 10 years [1, 3].

The three major forms of MPGN are identified by microscopy and immunofluorescence staining, and are differentiated due to their distinct morphology, the site of damage or deposit formation. The most common variant is MPGN type I (MPGN I, also called mesangiocapillary glomerulonephritis) in which deposits are formed within the subendothelial layer of the glomerular membrane. MPGN type II (MPGN II, also called dense deposit disease) is associated with deposits in the GBM. The term MPGN type III is used when deposits are formed in the subepithelial zone.

Type I membranoproliferative (mesangiocapillary) glomerulonephritis

In MPGN I, periodic acid-Schiff (PAS) staining normally shows doubling or complex replication of the basement membrane, a thickening of the capillary walls and mesangial matrix expansion. Hypercellularity can be explained by mesangial cell proliferation and capillary wall thickening, which may cause hypersegmentation. The glomerular membrane disruption causes a change in urine filtration, making the glomerulus permeable to protein and blood cells. In many patients with MPGN I, C3 is identified as a component in the deposits, especially in the idiopathic childhood variant.

Type II membranoproliferative glomerulonephritis

MPGN II is a rare disease and is characterized by the deposition of electron-dense material within the GBM of the kidney. MPGN II is considered a non-immune complex-mediated glomerulonephritis. Frequently, hypocomplementemia is associated with this disease, and the hyperactive C3 catabolism can be caused by inherited or acquired deficiency of the complement inhibitor Factor H, or by an autoantibody termed C3 nephritic factor (C3NeF), which binds to and stabilizes the alternative pathway convertase C3bBb [4].

Electron microscopy is useful for differentiating between type I and type II MPGN. In MPGN II, there is hypercellularity and thickening of capillary walls [5]. Some patients with this disease have thick capillary walls, but no hypercellularity, therefore the term dense deposit disease is frequently used. Histologically stained sections show a thickening of the basement membrane and also of the capillary wall. In most cases the dense deposits form within the basement membrane, but rarely in the subepithelial or subendothelial region. Immunofluorescence microscopy shows normally intense staining for C3, with almost no staining for immunoglobulins. The capillary wall staining is linear or bilinear. Electron microscopy often reveals spherical or ring-shaped mesangial dense deposits. MPGN II has a higher frequency of hypocomplementemia and of the presence of C3NeF than MPGN I.

It is difficult to differentiate between MPGN I and MPGN II by light and immunofluorescence microscopy. Immunofluorescence glomerular staining for C3 and lack of staining for immunoglobulin are common for MPGN II, but have also reported for MPGN I. Similarly hypocomplementemia is common for MPGN II and is also observed in MPGN I.

Based on the initial diagnosis of the various Factor H-deficient cases, several patients with defective Factor H were initially diagnosed as MPGN I and/or MPGN II. The reasons for this difference in diagnosis are currently unclear. They may reflect different disease states, or various time from the initial onset, they might even correlate with the type of Factor H gene mutation or reflect difficulties in the diagnosis.

The alternative pathway as an essential part of complement

Complement is part of innate immunity and represents a central defense system of the vertebrate host [6–8]. This cascade system is activated by three pathways and functions in:
(i) the elimination of foreign microbes or modified self cells
(ii) induction of a potent inflammatory response
(iii) the enhancement of the adaptive immune reactions
(iv) the clearance of immune complexes.

The alternative pathway of complement is the most ancient pathway, which is aimed at identifying, recognizing and eliminating microbes or modified tissue cells. Upon entry of a microbe into an immunocompetent host, the alternative pathway is directly initiated and amplified.

The activated complement system has disastrous effects, which are desirable on foreign surfaces, such as microbes, but are highly undesirable for host cells. The effector functions of the activated complement system include the generation and release of the potent anaphylatoxins C3a and C5a. The activated complement system causes surface deposition of a large number of C3b molecules. This process is termed opsonization, as it facilitates and enhances phagocytosis of the target particle. In addition, the activated system forms the terminal complement membrane attack complex, C5b-9 (MAC). Host integrity requires tight control and continuous downregulation of these toxic reactions, and when left uncontrolled damage also occurs to host cells and results in autoimmune diseases.

Default setting and activation: formation of the amplification convertase

The alternative pathway of complement represents a true safeguard system of the human host. Its central component C3 is found in high concentration in the plasma (1.4 mg/mL) and is distributed with blood and body fluids. Consequently, this system acts everywhere [6].

The initiation reaction—the initial phase

C3 initiates the alternative pathway by undergoing a spontaneous conformational change. This transient form of C3 is termed C3(H_2O), is generated in the fluid phase at a low level, has a very short half-life of milliseconds and is highly reactive. Upon binding of Factor B, the C3(H_2O)B complex is cleaved by Factor D. The Ba fragment is released, and the C3(H_2O)Bb complex forms the first and initial C3 convertase. These initiation reactions are rather slow, but trigger the alternative pathway and set the amplification cascade in motion (Fig. 1) [9].

The amplification reaction—the second phase

The initial and first convertase is formed in fluid phase and can activate C3. Upon cleavage of C3, the soluble anaphylatoxin C3a is generated and, in the newly formed C3b molecule, the highly reactive thioester is accessible. This initial highly reactive form of C3b* has an exposed thioester that immediately reacts with any molecule in its neighborhood, and binds preferentially to surfaces. The surface-bound C3b molecule can bind Factor B and upon cleavage by Factor D forms the surface-bound amplification convertase of the alternative pathway C3bBb. This active enzyme generates more C3b, and C3b is deposited to the surface and the number of new amplification convertases expands explosively. Thus, a potent amplification loop is generated, which acts like a chain reaction, and which can locally deposit around 10^6–10^8 C3b molecules at the site of action within 5–15 min [8, 10]. If activation proceeds, an active C5 convertase C3bBbC5b is generated. This convertase pathway initiates the terminal complement and forms the terminal complement complex C5b-C9, which inserts into membranes and forms pores, causing cell lysis.

In its default setting, i.e., when left uncontrolled, activation proceeds unrestricted, and the cascade expands explosively (Fig. 1). Due to its indiscriminatory nature, an activated complement forms a local toxic environment on any surface and causes damage. This kind of activation and damage is favorable on foreign surfaces and results in the elimination of a microbe. However, on host cells the same response is highly unfavorable. Therefore, host cells need to actively and continuously down-regulate the amplification cascade, particularly during the initial phase. Host cells use membrane-integrated inhibitors and attach plasma inhibitors to their surface to efficiently and competently inactivate the convertase and inhibit this process.

Regulation and regulators

The activity of the amplification convertase C3bBb is tightly controlled both in the fluid phase and on the surface of host cells or tissues. Multiple regulators exist, all of which prevent formation of the alternative pathway amplification convertase and which directly inactivate the convertase. The soluble regulators Factor H and the Factor H-like protein 1 (FHL-1), an alternatively spliced product of the human Factor H gene, are present in plasma and body fluids [11, 12]. The two regulators control complement in fluid phase and, in addition, attach to cells, where at the cell surface they form a second layer of control [11, 13]. The three integral membrane regulators complement receptor 1 (CR1, CD35), membrane co-factor protein (MCP, CD46) and decay-accelerating factor (DAF, CD55) are expressed on the surface of host cells [14–16].

The five human regulators have overlapping and redundant activities, and they:

(i) control and inhibit formation and assembly of the initial and the amplification C3 convertase. Thus, in the presence of a single regulator no amplification convertase is formed.

(ii) inactivate the newly generated surface-deposited C3b molecule. The regulators display cofactor activity for the serine protease Factor I. Factor I inactivates and cleaves a newly generated C3b protein and requires an appropriate cofactor protein. The two fluid-phase regulators Factor H, FHL-1 and the membrane-bound inhibitors CR1 and MCP share cofactor activity.

(iii) destroy an existing C3bBb convertase. The two fluid-phase regulators Factor H, FHL-1, and the two membrane-inserted regulators CR1 and DAF have decay-accelerating activity. Each protein individually accelerates the decay of the convertase complex.

Site of action and regulation

The alternative pathway convertase C3bBb is generated spontaneously and everywhere in the body. Due to the toxic effects of the activated complement system, the active convertase needs to be controlled anywhere and at any time. Host cells and most host tissue surfaces are equipped with use membrane-bound regulators and in addition surface-attached regulators to control this step.

The five host regulators show wide distribution and each cell or tissue utilizes a unique pattern and a different combination (mixture) of regulators. Some host cells

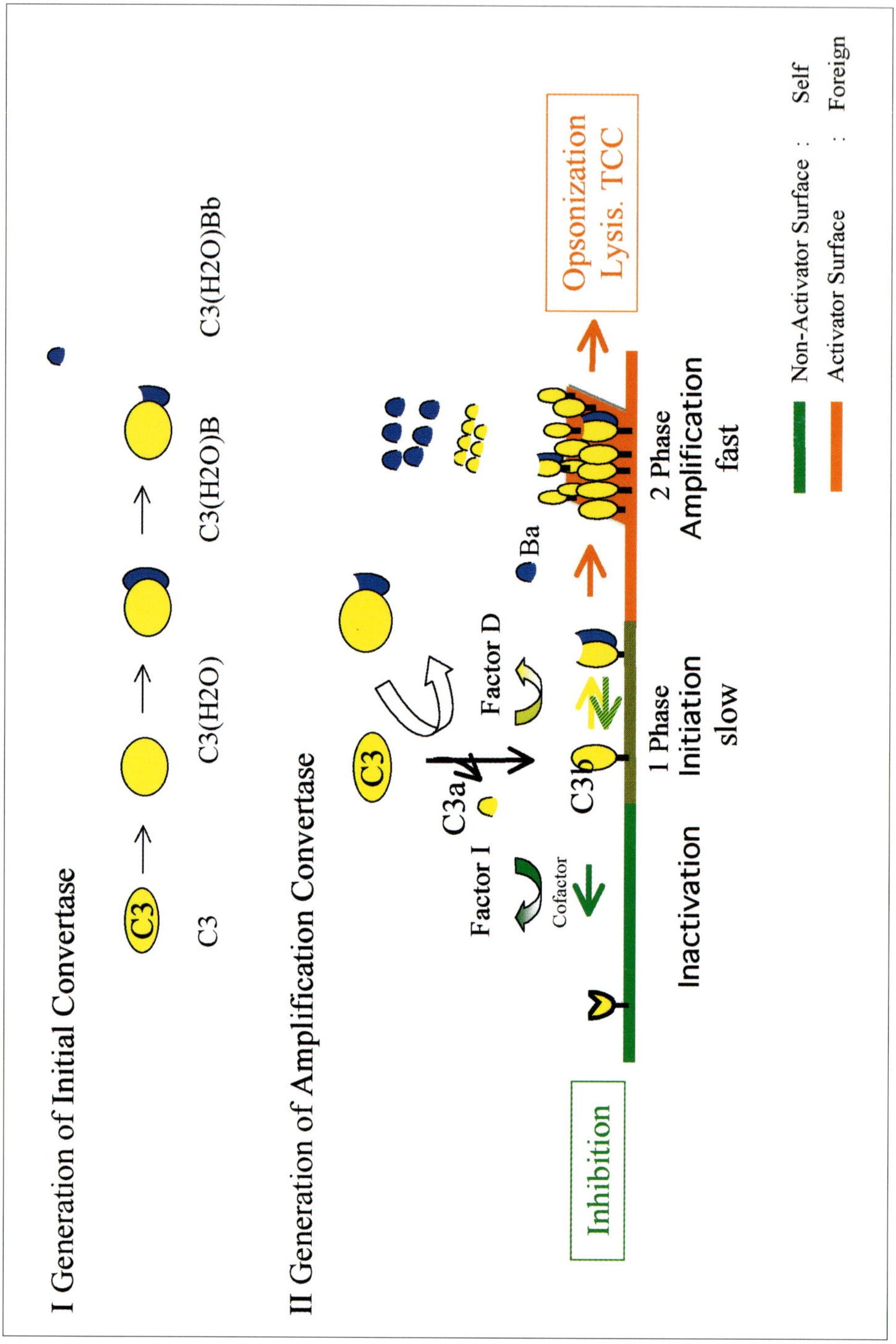
I Generation of Initial Convertase
C3
C3(H2O)
C3(H2O)B
C3(H2O)Bb
II Generation of Amplification Convertase
C3
C3a
Factor I
Cofactor
Factor D
C3b
Ba
Inhibition
Inactivation
1 Phase
Initiation
slow
2 Phase
Amplification
fast
Opsonization
Lysis. TCC
Non-Activator Surface : Self
Activator Surface : Foreign

express a single membrane-integrated regulator in high copy number, other cells express several integral regulators and also attach soluble fluid-phase regulators to their surface. In addition, some tissues lack endogenous regulators and dependent exclusively on soluble regulators.

Role of Factor H

Factor H and FHL-1 are soluble regulators distributed with plasma and body fluids that inhibit complement activation. Recent work has identified a role of Factor H as a cell surface binding protein. Factor H binds to endothelial cells via its C-terminal recognition site. When attached to the cell surface Factor H forms a second, additional layer that controls complement activation [13, 17]. In addition, for immune evasion, pathogenic microbes mimic host surface characteristics and express unique surface molecules for acquisition of host complement regulators Factor H and FHL-1 [18].

Expression of complement regulators on kidney cells and surfaces

Kidney cells express integral membrane-inserted complement regulators and, in addition, some cells also attach soluble complement inhibitors to their surface

Figure 1
Complement activation via the alternative pathway
Alternative complement activation is initiated in the fluid phase by the spontaneous conformational change of the component C3. The resulting protein C3(H_2O) binds Factor B, and, upon cleavage by Factor D, forms the initial fluid-phase alternative pathway convertase C3(H_2O)Bb. This initial convertase can convert fluid-phase C3 to an active C3b molecule, which has a highly reactive thioester exposed. This exposed thioester can immediately form covalent bounds with any moiety in its direct neighborhood. In addition, the soluble anaphylactic peptide C3a is generated and released. Bottom panel: The activated C3b* molecule can interact and bind to surface molecules. Following covalent attachment Factor B can bind and, upon cleavage by Factor D, the Ba fragment of Factor B is released and the C3bBb molecule forms the active alternative pathway amplification convertase. This molecule initiates the amplification cascade and generates further C3b molecules in its vicinity, which attach to the surface in direct vicinity of the enzyme and can form additional active enzymes, which perpetuate the amplification of the complement cascade. The activated system results in deposition of a large number of C3b molecules to the surface, (a process termed opsonization) or initiates the terminal complement reactions that forms the MAC or C5b-C9. See also the text for details.*

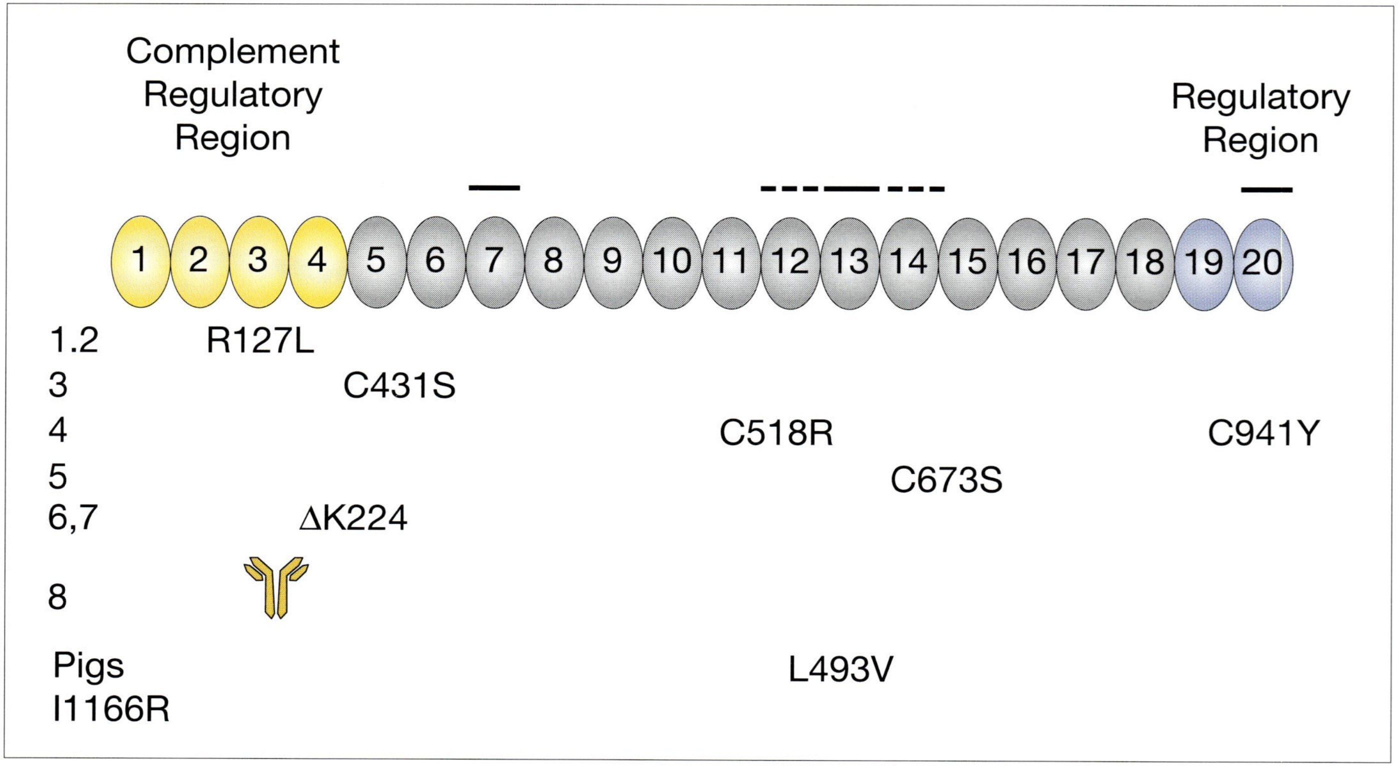

Figure 2
Structure of Factor H and localization of the reported Factor H gene mutations leading to MPGN
Factor H is organized in 20 consecutive domains termed complement control protein modules (CCPs) or short consensus repeats. The N-terminal complement regulatory domain of Factor H is contained in the first 4 CCP domains and is shown in yellow and the C-terminal recognition domain, that interacts with cell surface structures are contained in CCPs 19, 20 are shown in green.

Table 1 - Expression of complement regulators in kidney cells

	Complement regulator					
	Soluble		Membrane inserted			
Cell type	Factor H	FHL-1	CR1 (CD35)	MCP (CD46)	DAF (CD55)	Protectin CD59
Endothelial cells	ND	ND	–	+	+	+
Mesangial cells	+	ND	–	+	+	+
Epithelial cells (podocytes)		ND	+	+	+	+
Proximal tubular cells	+	ND	–	+	+	+/–
Distal tubular cells		ND	–	+	+	+
Collecting duct		ND	–	+	+	+

(Tab. 1). Endothelial and mesangial cells express MCP and DAF, but do not stain positive for CR1 [19, 20]. Epithelial cells, such as the podocytes express a combination of CR1, MCP, DAF and CD59. In addition, kidney cells also express and secrete soluble complement inhibitors; mesangial cells and proximal tubular cells secrete the soluble regulator Factor H. For cultured glomerular epithelial cells and glomeruli, expression of Factor H has been shown at both the mRNA and the protein level [21]. Expression is upregulated in membranous nephropathy [22]. Thus, it is likely that glomerular cells respond to complement attack and inflammation by inducing Factor H expression as a protective response. The newly secreted Factor H protein acts in an autocrine fashion; upon secretion the newly formed inhibitor binds directly to the secreting and to neighboring cells.

The existence, different distribution and simultaneous expression of several regulators explains why, under normal conditions, the absence or functional inactivation of a single regulator can be tolerated at most sites. The remaining proteins which display redundant activity can in combination control the activated complement system at a local site.

The GBM of the kidney is unique in its composition and structure, and is distinct form the adjacent endothelial and epithelial cells, as it lacks endogenous membrane-inserted regulators [23]. This tissue has a highly negatively charged surface, which binds, attaches, or absorbs regulators from plasma. Apparently, Factor H, but not the FHL-1, binds to the GBM, and the attached regulator is essential for local complement control.

This scenario or set up explains why either the absence of Factor H or a deregulated amplification convertase causes local damage, specifically of the GBM. As the adjacent endothelial cells and epithelial podocytes express membrane-bound regulators, it is hypothesized that these membrane-inserted regulators are sufficient to inhibit local complement control at the surface of these cells.

Dysregulation of the alternative pathway results in membranoproliferative glomerulonephritis

MPGN and hypocomplementemia has been reported in patients who:

(i) lack Factor H in plasma
(ii) express a defective Factor H protein
(iii) have a soluble inhibitor of Factor H
(iv) are positive for an autoantibody to the amplification convertase C3NeF.

Hypocomplementemia due to continuous C3 degradation is evidenced in plasma by low APH50 and CH50 values, low C3 and low Factor B levels, and the appearance of C3 degradation product C3d. Thus, it is hypothesized that a defective regulation of the amplification convertase causes pathophysiology [24].

Defective Factor H expression or deregulated function of Factor H

Complement regulation is essential and a proper control of the amplification convertase is particularly essential for integrity of kidney cells and particularly the basement membrane of the kidney.

Factor H gene mutations associated with MPGN

Factor H deficiencies have been reported in 29 individuals from 12 families ([25–32], reviewed in [33]); in addition, 2 patients from 1 family have been reported which express Factor H in their plasma [34]. So far, genetic Factor H mutations have been reported in a relative small number of patients; however, the emerging data allow a better and detailed insight in the mechanisms of the disease. Genetic defects in the Factor H gene of patients with MPGN II have been reported in 7 patients in 5 families and in two animal models (Fig. 2), Factor H-deficient pigs and Factor H-knockout mice provide valuable systems to analyze the pathomechanisms of the disease (Tab. 2). Both alleles of the Factor H gene are affected in MPGN patients. The patients may, therefore, represent a homozygous or a compound heterozygous scenario. This is further confirmed by the fact that heterozygous knockout mice are healthy and do not develop MPGN (see below).

Table 2 - Factor H gene mutations in patients with MPGN

Patient	Diagnosis	Factor H in plasma	CH50	C3	Amino acid	Domain	Comment	Refs
1, 2	MPGN	Absent	<10	low	R127L	CCP 2	2 patients, 1 family	[35]
3	MPGN I	Absent	<10	low	C431 S	CCP 7		[35]
4	MPGN I collagen type III glomerulopathy	Absent	nd	nd low	C518 R C941Y	CCP 9 CCP16	FHL-1 present	[36]
5	MPGN I nephrotic syndrome	Absent	<10	low	C637 S	CCP 11		[35]
6, 7	MPGN II	Present	<10	nd	Delta K227	CCP 4	2 patients, 1 family	[34]
8	MPGN I and MPGN II	Present	nd	nd	Auto-antibody	CCP 3		[39]
Pigs	MPGN II	Absent	nd	nd	L493V I1166R	CCP 9 CCP 20		[38]
Mice knockout	MPGN II	Absent	nd	nd	Knockout			[42]

Factor H gene mutations observed in patients with defective secretion

Cases 1 and 2

Two brothers with MPGN show a homozygous mutations in complement control protein module (CCP) 2 of the Factor H protein, which has the non-framework arginine residue at position 127 changed to leucine (R127L). At the first diagnosis the patients were 12 and 14 years old; Factor H was undetectable in their plasma, hypocomplementemia and complement activation were evidenced by low CH50 value (< 10% in both patients), low C3 (170 and 100 mg/L) and low Factor B plasma levels (70 and 50 mg/L) [35].

Case 3

Another patient, originally diagnosed with MPGN I and nephritic syndrome, shows homozygous exchange of the framework cysteine residue 431 to serine (C431S). This patient was 6 years old at the initial diagnosis, and plasma of this patient lacked Factor H, and complement activation was evidenced by low CH50 (<10) and low C3 (< 40 mg/L) and low Factor B levels (12 mg/L) [35].

Case 4

A 13-month-old native American patient with Factor H deficiency and hypocomplementemic hypertensive renal diseases was initially diagnosed as collagen type III glomerulopathy. Renal biopsy showed changes consistent with MPGN deposition of type III collagen and segmental complement C3 deposition in capillary loops [32]. In plasma, Factor H was absent and C3 and Factor B levels were decreased. Genetic analyses revealed a compound heterozygous deficiency with mutations of conserved cysteine residues in CCP 9 (C518R) and CCP 16 (C991Y). Both mutations caused a profound block in protein secretion. The mutant protein was expressed intracellularly, it was identified in the cytoplasm, but secretion across the cell membrane was blocked. The mutant protein accumulated in the endoplasmic reticulum and degraded relatively slowly. This patient expressed the related FHL-1 at about a 2.5-fold-increased level. Although Factor H and FHL-1 have similar functions *in vitro*, at least in this patient the FHL-1 protein did not substitute for the defective Factor H protein [36, 37].

Case 5

An additional patient diagnosed with MPGN I and nephritic syndrome also showed a homozygous mutation of the Factor H gene, which within CCP domain 11 converted a framework cysteine to a serine residue (C637S). Factor H was absent in the plasma of this patient and complement activation was indicated by low CH50 levels (< 10), low C3 (<40 mg/L) and low Factor B levels (17 mg/L) [35].

A comparison of the available data shows that a defective Factor H secretion is caused by mutations of central framework cysteine residues. Disulfide bond formation is required for the proper folding of an individual CCP domain, thus explaining why the lack of one single cysteine residue results in protein misfolding, the mutant protein accumulates in the Golgi complex and causes defective secretion. As indicated for patients 1 and 2 and for the deficient pigs, mutations of non-framework residues, such as R127L and I1166R, also result in defective secretion [35, 38].

Mutations that result in a secreted but functionally defective protein

Cases 6 and 7

Two patients, diagnosed as having MPGN II, who come from one family, expressed

Factor H in plasma. Complement activation was evidenced by low APH50 and CH50 plasma levels, low values for C3b and Factor B, and increased values of C3 degradation product C3d. Genetic analysis of the Factor H gene revealed the same homozygous deletion of a lysine residue at position 224 (K_{224}) in both patients. This deleted amino acid residue is located within the complement regulatory region in CCP 4, and is relevant for pathophysiology. The mutant Factor H protein was purified from plasma from both patients and showed defective complement regulatory activity, i.e., severely reduced cofactor, as well as reduced decay-accelerating activities and reduced binding to the central complement component C3b. Regular infusion of fresh frozen plasma (FFP) stopped disease progression in both patients and preserved kidney function. To our knowledge, this study reports the first mutant Factor H protein expressed in plasma of two MPGN II patients that displays defective complement control [34].

Autoantibodies that inactivate the regulatory function of Factor H

Case 8

Autoantibodies that bind to Factor H and that inhibit the regulatory functions of the protein represent an additional cause for MPGN. In a 57-year-old woman who developed hypocomplementemia and MPGN, extensive glomerular deposits of C3 were observed in the kidney and mesangial matrix and thickening of the capillary walls was evident. The GBM was thickened and showed dense intramembranous subendothelial deposits. Immunohistology analyses of the patient's kidney revealed an atypical form of MPGN with both subendothelial (type I) and dense intramembranous deposits (type II) [39]. Subsequently, a circulating factor was purified from the plasma and urine of the patient, which was identified as a 46-kDa monoclonal immunoglobulin lambda light chain dimer. This lambda light chain activated the alternative pathway in a dose- and ion-dependent manner, and bound to Factor H, but not to the amplification convertase C3bBb. The binding site of the lambda light chain was localized to CCP 3 of Factor H, which represents the regulatory region of the protein. Binding of the autoantibody inhibited cofactor and decay acceleration activity. Binding of this autoantibody blocked Factor H binding to C3b, and thus, by inhibiting the complement regulator, caused uncontrolled activation of the alternative pathway [40].

Animal models

Two animal models for MPGN II have been reported: Factor H-deficient pigs represent natural mutants [41] and Factor H-knockout mice were genetically designed

[42]. The pigs and mice have provided, and continue to provide valuables systems for analyzing the pathomechanisms of MPGN II.

Factor H-deficient pigs

An initial report showed a significant loss among Norwegian Yorkshire pigs due to kidney defects [42]. The pigs developed MPGN II (dense deposit disease) early in life and died at about 1–2 month after birth. Glomerular disease was evidenced by deposition of complement components, which began *in utero*, such as C3b and terminal complement complex (TCC) in the GBM [41, 43]. Serum analyses showed the absence of Factor H in plasma, and genetic analyses identified homozygous mutations of non-framework residues L493V and I1166R that are located within CCP 9 and CCP 20 [38]. The mutant protein is expressed, accumulates intracellularly and is absent in plasma.

A detailed analysis combining light microscopy, immunohistology and electron microscopy followed the alterations in the glomeruli of the kidneys of the animals in a chronological manner [44]. C3 and TCC staining of the glomeruli were identified already 4 days after birth and the intensity of staining increased with time. C3 deposition preceded the appearance of subendothelial electron-dense deposits and the structural changes of the GBM. The process starts at the subendothelial site, the site being exposed to Factor H-deficient plasma. Heterozygous pigs, which show half normal plasma concentration of Factor H are healthy and are indistinguishable from wild-type animals.

Factor H-knockout mice

Factor H-knockout mice provide a valuable model system for studying the pathomechanisms of MPGN [42]. Factor H-deficient mice were generated on a C57BL/6 background. Factor H is absent in plasma of these mice and the animals develop MPGN II, relatively late in life. At 8 month of age the knockout mice display a mortality of 25%; thus, mice are older than the pigs when they develop the disease [45]. However, similar to deficient pigs, the knockout mice develop MPGN spontaneously and are hypersensitive for developing renal injury. These analyses confirmed the absolute requirement of uncontrolled C3 activation in the pathogenesis of MPGN. Uncontrolled C3 activation *in vivo* is associated with deficiency, or functionally inactivity of Factor H [42]. The knockout animals show reduced levels of C3 in plasma. When assayed at 4 days of age, there was deposition of C3 and C9, seen by immunofluorescence, in capillary wall of the glomeruli, but histology of the kidney was normal. At 2 month of age, C3 and C9 were still detectable and at this time linear subendothelial electron-dense deposits were present in the capillary wall. These alterations demonstrate that uncontrolled complement activation leads to deposition of C3 and C9. These deposits are present and can be identified for a long period before glomerulonephritis develops. In addition, the Factor H-knockout mice show increased susceptibility to experimental nephritis.

C3 nephritic factor, an autoantibody that inactivates the alternative complement pathway convertase C3bBb

C3NeF is an autoantibody to the alternative pathway convertase C3bBb. It is a gamma globulin with an estimated molecular mass of 150 kDa, and there is (increasing) evidence for an autoantibody IgG nature of C3NeF. The native C3 convertase is rather labile and has an in vivo half-life of 4 min. C3NeF stabilizes the C3bBb complex by protecting it from Factor I inactivation. Binding of C3NeF increases the half-life of the enzyme about tenfold to 45–60 min [46–49]. As a consequence a C3NeF-stabilized amplification convertase consumes more C3b, resulting in amplified C3 cleavage in serum, enhanced local complement activation and increased C3 deposition.

Between 54% of adult and 82% of pediatric MPGN patients are positive for C3NeF, which binds to the C3bBb. The presence of C3NeF is associated with MPGN, similar as Factor H deficiency. However, C3NeF is also observed in patients with partial lipid dystrophy, menigococcal meningitis, and anti-phospholipid syndrome, and even in healthy individuals, who show no clinical signs of glomerulonephritis [50–53].

The reported examples demonstrate that distinct scenarios, which all affect the amplification convertase of the alternative pathway of complement, lead to MPGN. In addition, based on the understanding of the situation in hemolytic uremic syndrome (see chapters by Skerka and Józsi, and Goodship et al.), it is hypothesized that additional types of mutation in the Factor H gene exist in patients with MPGN. If these mutations occur in a homozygous scenario, they will also lead to MPGN:

- (i) point mutation in the Factor H gene that cause misfolding and a block of protein secretion and result in the absence of Factor H in plasma (cases 1–5)
- (ii) point mutations that affect central functions of the secreted protein. e.g., the regulatory N-terminal region (cases 6 and 7)
- (iii) autoantibodies to Factor H that bind to central functional domains, i.e., the N-terminal regulatory and the C-terminal recognition domain (case 8)
- (iv) nonsense mutations that introduce a premature stop codon and cause a block of protein synthesis are likely to exist and cause the same effect (not shown so far)
- (v) autoantibodies, termed C3NeF that bind to the C3 convertase and thereby stabilize the enzyme and increase enzymatic activity.

All scenarios affect or modify the activity of the alternative pathway amplification convertase and consequently increase the activity of this enzyme and result in pathophysiology.

Pathomechanisms

Quite distinct scenarios that all cause dysregulation of the amplification convertase lead to glomerulonephritis. Deregulation of the amplification convertase of complement occurs systemically and results in excessive C3b consumption. This systemic effect is evidenced in plasma by low APH50 activity, reduced C3 and Factor B plasma levels, and by the generation of C3 degradation product C3d.

Although the deregulation is systemic and occurs everywhere in the body, the pathological changes occur locally and appear specifically at the GBM. This local damage is explained by the unique structure and specific composition of the GBM. This membrane lacks endogenous membrane-inserted regulators and requires exclusively Factor H for protection. This requirement is distinct from that of directly adjacent endothelial and epithelial cells, which express membrane-bound regulators. If Factor H is absent, and as the GBM depends exclusively on Factor H, defects in form of local complement activation, induction of the amplification convertase and C3b deposition occur specifically at the site of the GBM. In contrast, as the adjacent endothelial and epithelial cells express membrane-integrated regulators, activation is controlled and stopped at the cell surface. As has been shown in the Factor H-deficient pigs complement deposition starts at the subendothelial side of the GBM, which is the side being exposed to blood through the fenestrated endothelial of the glomerular capillaries.

The structure and composition of the GBM provide further clues for the damaging effects [54]. Once complement is activated on a cell, apoptosis, phagocytosis or lysis is induced and the damaged host cell is eliminated and consequently disappears. However, when complement activation proceeds on a tissue surface, like the basement membrane, this surface can not undergo apoptosis or lysis. Consequently, these structures can neither be removed by phagocytosis nor can lysis or MAC formation occur. Under these conditions a defective control results in continuous, local, complement activation and the process becomes chronic. Immunofluorescence analyses of the dense deposits, which stain positive for C3b and MAC are thus indicative of a chronic complement attack. It is currently unclear how the glomerular membrane responds to this chronic attack, which last for months or even years. Most likely repair processes are induced, which, in the situation of continuous complement activation and C3 deposition, result in interruptions of the glomerular membrane, which explains the gap formation that develops into the double contour structures.

It has been known that patients with defective complement are at risk to develop age-related macular degeneration [55]. Extracellular deposits are formed between the retinal pigmented epithelium and choroids. Immunohistological analyses identified several complement components such as C5 and C5b-9 complexes in all type of drusen, thus indicating a defective complement control at this site [56]. Recent genetic data have shown that a previously identified polymorphism of the Factor H gene, which causes a H402Y exchange, is strongly associated with age-related macular degeneration [57–59]. These recent results from the human genome showed that eye diseases are also caused by defective complement control.

Outlook: diagnosis, treatment and prognosis for transplanted kidneys

Diagnosis

Activation of complement in patients can be monitored as a depressed total complement hemolytic activity, indicated as APH50 or CH50. Decreased levels of individual proteins, which are consumed during the continuous activation process, e.g., C3 and Factor B, and similarly the presence of activation split products, e.g., C3a and C3d, can be directly assayed in plasma, and are clear indication for complement activation. Further, the level of the individual regulators such as Factor H, or of the serine protease Factor I and the levels of autoantibodies for the individual regulators and for C3NeF should be assayed in plasma (see chapters by Noris and Remuzzi, Licht and Hoppe, and Würzner and Zimmerhackl). The presence of membrane-bound regulators such as MCP and DAF can be assayed in blood cells by means of flow cytometry. Although these assays and plasma tests are valuable tools, they do not cover all mutations as, e.g., normal serum levels for C3 and for Factor H do not exclude functional defects. Therefore, genetic testing for the various relevant genes should be considered in addition (see also chapter by Goodship et al.).

Treatment

In several cases, a deregulation of the complement system caused by a defective or missing regulator such as Factor H is treated by plasma infusion and/or exchange. In a Factor H-deficient patient the half-life of Factor H was shown to be 6.5 days, thus giving a rationale for plasma administration in 2-week intervals [61]. In most cases, regular plasma infusion are well tolerated. In addition, it has recently been suggested that the complement system of these patients could be modulated using complement inhibitors [61, 62] (see chapters by Licht and Hoppe, and Würzner and Zimmerhackl).

Prognosis for transplanted kidneys

In case of defective systemic complement deregulation, there is a poor prognosis for a kidney transplant. Two recent reviews show that patients with Factor H deficiency have a high risk for diseases recurrence, and in addition a high frequency for loss of the grafted kidney [63, 64]. This situation is different for patients, who show mutations of integral membrane regulators such as MCP: upon transplantation the defect is repaired, as the transplant expresses an intact regulator. Therefore, this situation is likely to exclude diseases recurrence. In addition, the combined liver and kidney transplantation that have been reported recently [65, 66] have been unsuccessful in their long-term outcome.

Conclusions

The alternative pathway of complement mediates immune defense and regulates tissue integrity. The recent characterization of the role of this complement activation pathway in distinct diseases, such as MPGN, atypical form of hemolytic uremic syndrome, age-related macular degeneration of the eye and in microbial immune evasion, highlight the importance of this branch of complement for the integrity of the human body. Proper regulation maintains tissue integrity, and improper regulation results in local defects which eventuate in severe diseases. The detailed understanding of the individual reactions and of the local role of the individual regulators defines how this part of complement maintains tissue integrity and how defects result in diseases. This understanding is essential to define proper tools and approaches for diagnosis and all kind of treatments which modulate the action of the complement system systemically and locally.

Acknowledgement
The work of the authors is funded by the Deutsche Forschungsgemeinschaft, the Kidneeds program, Cedar Rapids, IA, USA and the Foundation for Children with Atypical HUS.

References

1 Davison AM, Cameron JS, Grünfeld JP et al (2005) *Oxford Textbook of Clinical Nephrology*. Oxford University Press, Oxford
2 Appel GB, Cook HT, Hageman G, Jennette JC, Kashgarian M, Kirschfink M, Lambris

JD, Lanning L, Lutz HU, Meri S et al (2005) Membranoproliferative glomerulonephritis Type II (dense deposit disease): an update. *J Am Soc Nephrol* 16: 1392–13403
3 Orth SR, Ritz E (1998) The nephrotic syndrome. *N Engl J Med* 338: 1202–1211
4 Habib R, Gubler MC, Loriat C, Maiz HB, Levy M (1975) Dense deposit disease. A variant of membranoproliferative glomerulonephritis. *Kidney Int* 7: 204–215
5 Cameron JS, Turner Dr, Heaton J, Gwyn Williams D, Ogg CS, Chantler C, Haycock GB, Hicks J (1983) Idiopathic mesangiocapillary glomerulonephritis. Comparison of types I and II in children and adults and long-term prognosis. *Am J Med* 74: 175–192
6 Walport MJ (2001) Complement. First of two parts. *N Engl J Med* 344: 1058–1066
7 Walport MJ (2001) Complement. Second of two parts. *N Engl J Med* 344: 1140–1144
8 Muller Eberhard HJ (1988) Molecular organization and function of the complement system. *Annu Rev Biochem* 57: 321–347
9 Pangburn MK (1998) Alternative Pathway: Activation and Regulation. In: K Rother, GO Till, GM Hänsch (eds): *The Complement System*. Springer, Berlin, 93–115
10 Pangburn MK, Muller-Eberhard HJ (1986) The C3 convertase of the alternative pathway of human complement. *Biochem J* 235: 723–730
11 Zipfel PF, Jokiranta TS, Hellwage J, Koistinen V, Meri S (1999) The factor H protein family. *Immunopharmacology* 42: 53–60
12 Rodríguez de Córdoba S, Esparza-Gordillo J, Goicoechea de Jorge E, Lopez-Trascasa M, Sánchez-Corral P (2004) The human complement factor H: functional roles, genetic variations and disease associations. *Molecular Immunol* 41: 355–367
13 Józsi M, Manuelian T, Heinen S, Oppermann M, Zipfel PF (2004) Attachment of the soluble complement regulator factor H to cell and tissue surfaces: relevance for pathology. *Histol Histopathol* 19: 251–258
14 Hourcade D, Liszewski MK, Krych-Goldberg M, Atkinson JP (2000) Functional domains, structural variations and pathogen interactions of MCP, DAF and CR1. *Immunopharmacology* 49: 103–116
15 Krych-Goldberg M, Atkinson JP (2001) Structure-function relationships of complement receptor type 1. *Immunol Rev* 180: 112–122
16 Liszewski MK, Atkinson JP (1998) Regulatory proteins of complement. In: JE Volonakis, MM Frank (eds): *The Human Complement System in Health and Disease*. M. Dekker, Inc., New York
17 Manuelian T, Hellwage J, Meri S, Caprioli J, Noris M, Heinen S, Jozsi M, Neumann HP, Remuzzi G, Zipfel PF (2003 Mutations in factor H reduce binding affinity to C3b and heparin and surface attachment to endothelial cells in hemolytic uremic syndrome. *J Clin Invest* 111: 1181–1190
18 Zipfel, PF, Skerka C, Hellwage J, Jokiranta ST, Meri S, Brade V, Kraiczy P, Noris M, Remuzzi G (2002) Factor H family proteins: on complement, microbes and human diseases. *Biochem Soc Trans* 30: 971–978
19 van den Dobbelsteen ME, Verhasselt V, Kaashoek JG, Timmerman JJ, Schroeijers WE, Verweij CL, van der Woude FJ, van Es LA, Daha MR (1994) Regulation of C3 and fac-

tor H synthesis of human glomerular mesangial cells by IL-1 and interferon-gamma. *Clin Exp Immunol* 95: 173–180

20 Timmerman JJ, van der Woude FJ, van Gijlswijk-Janssen DJ, Verweij CL, van Es LA, Daha MR (1996) Differential expression of complement components in human fetal and adult kidneys. *Kidney Int* 49: 730–740

21 Nangaku M (1998) Complement regulatory proteins in glomerular diseases. *Kidney Int* 54: 1419–1428

22 Bao L, Spiller OB, St John PL, Haas M, Hack BK, Ren G, Cunningham PN, Doshi M, Abrahamson DR, Morgan BP, Quigg RJ (2002) Decay-accelerating factor expression in the rat kidney is restricted to the apical surface of podocytes. *Kidney Int* 62: 2010–2021

23 Pavenstädt H, Kriz W, Kretzler M (2003) Cell Biology of the glomerular podocyte. *Physiol Rev* 83: 253–307

24 Cunningham PN, Qigg RJ (2005) Contrasting role of complement activation and its regulation in membranous nephropahty. *J Am Soc Nephrol* 16: 1214–1222

25 Thompson RA, Winterborn MH (1981) Hypocomplementaemia due to a genetic deficiency of beta 1H globulin. *Clin Exp Immunol* 46: 110–119

26 Wyatt RJ, Julian BA, Weinstein A, Rothfield NF, McLean RH (1982) Partial H beta 1H) deficiency and glomerulonephritis in two families. *J Clin Immunol* 2: 110–117

27 Levy M, Halbwachs-Mecarelli L, Gubler MC, Kohout G, Bensenouci A, Niaudet P, Hauptmann G, Lesavre P (1986) H deficiency in two brothers with atypical dense intramembranous deposit disease. *Kidney Int* 30: 949–956

28 Nielsen HE, Christensen KC, Koch C, Thomsen BS, Heegaard NH, Tranum-Jensen HE (1989) Hereditary, complete deficiency of complement factor H associated with recurrent meningococcal disease. *Scand J Immunol* 30: 711–718

29 Rougier N, Kazatchkine MD, Rougier JP, Fremeaux-Bacchi V, Blouin J, Deschenes G, Soto B, Baudouin V, Pautard B, Proesmans W et al (1998) Human complement factor H deficiency associated with hemolytic uremic syndrome. *J Am Soc Nephrol* 9: 2318–2326

30 Fijen CA, Kuijper EJ, Te Bulte M, van de Heuvel MM, Holdrinet AC, Sim RB, Daha MR, Dankert J (1996). Heterozygous and homozygous factor H deficiency states in a Dutch family. *Clin Exp Immunol* 105: 511–516

31 Brai M, Misiano G, Maringhini S, Cutaja I, Hauptmann G (1988) Combined homozygous factor H and heterozygous C2 deficiency in an Italian family. *J Clin Immunol* 8: 50–56

32 Vogt BA, Wyatt RJ, Burke BA, Simonton SC, Kashtan CE (1995) Inherited factor H deficiency and collagen type III glomerulopathy. *Pediatr Nephrol* 9: 11–15

33 Kirschfink M, Binder R (1998) Deficiencies in the Alternative pathway: Factors I and H. In: K Rother, GO Till, GM Hänsch (eds): *The Complement System*. Springer, Berlin, 420–427

34 Licht C, Heinen S, Józsi M, Löschmann I, Saunders RE, Perkins SJ, Waldherr R, Skerka C, Kirschfink M, Hoppe B, Zipfel PF (2004) Deletion of a single amino acid (K224) in

the regulatory domain of the immune inhibitor Factor H reveals a novel pathomechanism for membranoproliferative glomerulonephritis type II. *Immunobiol* 209: 328

35 Dragon-Durey MA, Fremeaux-Bacchi V, Loirat C, Blouin J, Niaudet P, Deschenes G, Coppo P, Herman Fridman W, Weiss L (2004) Heterozygous and homozygous factor H deficiencies associated with hemolytic uremic syndrome or membranoproliferative glomerulonephritis: report and genetic analysis of 16 cases. *J Am Soc Nephrol* 15: 787–795

36 Ault BH, Schmidt BZ, Fowler NL, Kashtan CE, Ahmed AE, Vogt BA, Colten HR (1997) Human factor H deficiency. Mutations in framework cysteine residues and block in H protein secretion and intracellular catabolism. *J Biol Chem* 272: 25168–25175

37 Schmitt Schmidt BZ, Fowler NL, Hidvegi T, Perlmutter DH, Colten HR (1999) Disruption of disulfide bonds is responsible for impaired secretion in human complement factor H deficiency. *J Biol Chem* 274: 11782–11788

38 Hegasy GA, Manuelian T, Hogasen K, Jansen JH, Zipfel PF (2002) The molecular basis for hereditary porcine membranoproliferative glomerulonephritis type II. *Am J Pathol* 161: 2027–2034

39 Meri S, Koistinen V, Miettinen A, Tornroth T, Seppala IJ (1992) Activation of the alternative pathway of complement by monoclonal lambda light chains in membranoproliferative glomerulonephritis. *J Exp Med* 175: 939–950

40 Jokiranta TS, Solomon A, Pangburn MK, Zipfel PF, Meri S (1999) Nephritogenic l light chain dimer: A unique human miniautoantibody against complement factor H. *J Immunol* 163: 4590–4596

41 Hogasen K, Jansen JH (1995) Porcine membranoproliferative glomerulonephritis type II is caused by factor H deficiency. *J Clin Invest* 95: 1054–1061

42 Pickering MC, Cook HT, Warren J, Bygrave AER, Moss J, Walport MJ, Botto M (2002) Uncontrolled C3 activation causes membranoproliferative glomerulonephritis in mice deficient in complement factor H. *Nat Genet* 31: 424–428

43 Jansen JH, Hogasen K, Mollnes TE (1993) Extensive complement activation in hereditary porcine membranoproliferative glomerulonephritis type II (porcine dense deposit disease). *Am J Pathol* 143: 1356–1365

44 Jansen JH, Hogasen K, Harboe M, Hovig T (1998) In situ complement activation in porcine membranoproliferative glomerulonephritis type II. *Kidney Int* 53: 331–349

45 Alexander JJ, Pickering MC, Haas M, Osawe I, Quigg RJ (2005) Complement factor h limits immune complex deposition and prevents inflammation and scarring in glomeruli of mice with chronic serum sickness. *J Am Soc Nephrol* 16: 52–57

46 Daha MR, Van Es LA (1981) Stabilization of homologous and heterologous cell-bound amplification convertases, C3bBb, by C3 nephritic factor. *Immunol* 43: 33–38

47 Daha MR (1998) C3 Nephritic Factor. In: K Rother, GO Till, GM Hänsch (eds): *The Complement System*. Springer, Berlin, 456–462

48 Weiler JM, Daha MR, Austen KF, Fearon DT (1976) Control of the amplification convertase of complement by the plasma protein b1H. *Proc Natl Acad Sci USA* 73: 3268–3272

49 West CD, Witte DP, McAdams JA (2001) Composition of Nephritic Factor-Generated Glomerular Deposits in Membranoproliferative Glomerulunephritis Type 2. *Am J Kidney Dis* 37: 1120–1130

50 West CD (1994) Nephritic factors predispose to chronic glomerulonephritis. *Am J Kidney Dis* 24: 956–963

51 West C (1998) Complement and glomerular Diseases. In: JE Volonakis, MM Frank (eds): *The Human Complement System in Health and Disease*. M. Dekker, Inc., New York, 571–596

52 Mathieson PW, Wurzner R, Oliveria DB, Lachmann PJ, Peters DK (1993) Complement-mediated adipocyte lysis by nephritic factor sera. *J Exp Med* 177: 1827–1831

53 Schwertz R, Rother U, Anders D, Gretz N, Scharer K, Kirschfink M (2001) Complement analysis in children with idiopathic membranoproliferative glomerulonephritis: A long-term follow-up. *Pediatr Allergy Immunol* 12: 166–172

54 Mason RM, Wahab NA (2003) Extracellular matrix metabolism in diabetic nephropathy. *J Am Soc Nephrol* 14: 1358–1373

55 Leys A, Proesmans W, Van Damme-Lombaerts R, Van Damme B (1991) Specific eye fundus lesions in type II membranoproliferative glomerulonephritis. *Pediatr Nephrol* 5: 189–192

56 Mullins RF, Russell SR, Anderson DH, Hageman GS (2000) Drusen associated with aging and age-related macular degeneration contain proteins common to extracellular deposits associated with atherosclerosis, elastosis, amyloidosis, and dense deposit disease. *FASEB J* 14: 835–846

57 Klein RJ, Zeiss C, Chew EY, Tsai JY, Sackler RS, Haynes C, Henning AK, Sangiovanni JP, Mane SM, Mayne ST et al (2005) Complement Factor H polymorphism in age-related macular degeneration. *Science* 308: 385–389

58 Haines JL, Hauser MA, Schmidt S, Scott WK, Olson LM, Gallins P, Spencer KL, Kwan SY, Noureddine M, Gilbert JR, Schnetz-Boutaud N et al (2005) Complement Factor H variant increases the risk of age-related macular degeneration. *Science* 308: 419–421

59 Edwards AO, Ritter R 3rd, Abel KJ, Manning A, Panhuysen C, Farrer LA (2005) Complement Factor H polymorphism and age-related macular degeneration. *Science* 308: 421–424

60 Hageman GS, Anderson DH, Johnson LV, Hancox LS, Taiber AJ, Hardisty LI, Hageman JL, Stockman HA, Borchardt JD, Gehrs KM et al (2005) A common haplotype in the complement regulatory gene factor H (HF1/CFH) predisposes individuals to age-related macular degeneration. *Proc Natl Acad Sci USA* 102: 7227–7232

61 Licht C, Weyersberg A, Heinen S, Stapenhorst L, DevengeJ, Beck B, Waldherr R, Kirschfink M, Zipfel PF, Hoppe B (2005) Factor H deficiency and atypical hemolytic uremic syndrome successful plasma therapy for atypical hemolytic uremic syndrome (aHUS) caused by Factor H deficiency due to a mutation in the complement cofactor protein (CCP) domain 15. *Am J Kidney Dis* 45: 415–421

62 Gerber A, Kirchhoff-Moradpour AH, Obieglo S, Brandis M, Kirschfink M, Zipfel PF,

Goodship JA, Zimmerhackl, LB (2003) Successful (?) therapy of hemolytic-uremic syndrome with factor H abnormality. *Pediatr Nephrol* 18: 52–55
63 Loirat C, Niaudet P (2003) The risk of recurrence of hemolytic uremic syndrome after renal transplantation in children. *Pediatr Nephrol* 18: 1095–1101
64 Neumann HP, Salzmann M, Bohnert-Iwan B, Mannuelian T, Skerka C, Lenk D, Bender BU, Cybulla M, Riegler P, Konigsrainer A et al (2003) Haemolytic uraemic syndrome and mutations of the factor H gene: a registry-based study of German speaking countries. *J Med Genet* 40: 676–681
65 Remuzzi G, Ruggenenti P, Codazzi D, Noris M, Caprioli J, Locatelli G, Gridelli B (2002) Combined kidney and liver transplantation for familial haemolytic uraemic syndrome. *Lancet* 359: 1671–1672
66 Remuzzi G, Ruggenenti P, Colledan M, Gridelli B, Bertani A Bettinaglio P, Bucchioni S, Sonzogni A, Bonanomi E, Sonzogni V et al (2005) Hemolytic uremic syndrome: a fatal outcome after kidney and liver transplantation performed to correct factor H gene mutation. *Am J Transplant* 5: 1146–1150

The experience of a patient advocacy group

Pearl L. Lewis

Maryland Patient Advocacy Group, Foundation for Children with Atypical HUS, Maryland Renal Advocate 9131, 2530 Kensington Gardens #104, Ellicott City, MD 21043, USA

Introduction

What is a Patient Advocacy Group? It is so much more than the definition implies. The best of the best address patient, professional and public education; lobby on both the state and federal levels to assure citizens with the specific condition of access to medically necessary healthcare, raise funds for research and encourage promising researchers to address the disease to identify the cause, better treatment and cure for the condition.

The Foundation for Children with Atypical HUS

The Foundation for Children with Atypical HUS was founded in 2001, as most foundations are, by family members of someone with the disease. In this case the individual with the disease was the son of the founding person. With a disease as rare as this there was no where to turn for information or support. The founding family took the initial steps to create what exists today – a non-profit foundation that provides education, and support, and which has funded research projects in Italy and Germany (http://www.atypicalhus.50megs.com/). The Foundation for Children with Atypical HUS combines its efforts with The Nando Peretti Foundation (http://www.nandoperettifound.org) to support any activities and research that will lead to identifying the cause, and to better treatment and cure for atypical HUS.

Advocacy, education and support

In over 21 years as a patient advocate and interaction with thousands of patients, one thing is evident, patients who receive education and support from patient advocacy groups – comprised jointly of patients and physicians working together to encourage and foster a therapeutic alliance between physician and patient – lead

Complement and Kidney Disease, edited by Peter F. Zipfel

much fuller, healthier lives than those who have not. The cornerstone of this effort is to encourage and foster a therapeutic alliance between physician and patient. Patient advocacy groups have proven to be effective in (i) encouraging and raising funds for research; (ii) molding public policy; (iii) providing patient education and support, and (iv) educating the public.

Our world

The words of Albert Schweitzer best convey the impact of chronic illness as well as the belief in humanity's ability to hold fast to our dreams. "....life take(s) from us our belief in the good and the true and our enthusiasm for them. We need not surrender our dreams and ideals. Ideals, when brought into contact with reality, are for a time, crushed by facts. We, however, are not bound to capitulate to the facts, but merely to the fact that our ideals and dreams, for that moment, are on hold."

Those of us who live each day with chronic illness have had to wrestle with these fact innumerable times. If we are strong enough, introspective enough, we are able to put our lives on hold, and, when medical reality allows, we once again move forward reaching for the goals we have had to adjust time and again. Not everyone is able to manage this; in the wake of chronicity lives are crushed, dreams are shattered. Worldwide those with chronic illness, especially those with chronic and end-stage renal disease exist medical appointment to appointment, treatment to treatment. Time that was once filled with school, career, and family activities are now filled with painful imposing medical reality. What is it that can allow one to retain "normalcy" in the face of this new reality? Can one hold on alone?

While the power within is incalculable; something must happen to activate that power. It is available to all who seek it; it exists within the experience of those of us who have traveled the self-same path and devoted our lives to sharing what we know to be true, that there exists within us the strength to overcome that which afflicts our physical being. That is the role patient advocacy plays – we are there to extend our hand to those who are just beginning the journey we have successfully traveled.

The process of internalizing one's experience, coming to grips with that experience, and once doing so, transforming that knowledge into concrete intervention that will allow one to achieve optimal psychosocial function, is not an easy task. To develop a delivery system that will impart that knowledge to others is an even more difficult task, but such systems have been developed, they do exist, and they must be used in the care of patients with chronic disease.

Rabbi Harold Kushner, in his book *When Bad Things Happen to Good People*, shared his experiences with us as the father of a terminally ill child. In doing so, he has untangled for many parents the web of guilt, anger, and rage they feel. As a physician, you cannot explain the cause or cure for disease any more than you can explain life itself. You cannot tell parents why their child is ill, but as Kushner says, you can help them rise above the question "why" and move on to ask the question, "what do I do now that it has happened?" Stand back and for a moment put yourself in that parent's shoes. Imagine what that parent is feeling and will feel each and every day of his or her life.

Although the worst fear we envision is death for ourselves or our child, when kidney disease is at its worst this thought appears for a moment in the guise of "it might offer relief for us all." What kind of person or parents are we to allow this thought to even cross our minds? Our reality and permanent source of anxiety, this chronic, often unremitting entity, has taken hold of our child and family.

"The issue is not death; death is too final, too spiritual to grand. Chronic illness is an unsung presence that hovers, fluttering like an uncertain bird of prey. We are caught in a deep chasm between remote rims of acute illness and non-illness – there is no health. There are constant perturbations of sensation: these do not even deserve the word pain, but they are transformed by our imagination into agony.

There are waves of hope and despair with constant fatigue, irritability, and an endless preoccupation with numbers on lab reports and how one feels. There are waves of hope and despair with restless fingers probing for remembered aches. There is waiting in doctor's offices, laboratories, x-ray departments, drug stores, and for and in hospital beds. Health is less than a memory, more of a mirage – thin, unreal, longed for, elusive. Guilt runs rampant, "Why did I? Why didn't I?" How in the world can you smile while he/she/I am/is suffering?"

What fiscally driven medicine does at times is fragment this systemic disaster into fragments of internal organs, each of which is separately and uniquely treated with high technology and ineffective medicine, and in that reality that most precious gift is lost – humanity.

The physicians have the ability to interject humanity into that mix by making the connection between your patient and the education and support he/she so desperately needs. You cannot do it all. The patient whether parent or child needs sympathy and compassion, the reassurance that each is a good person, cherished as a worthwhile individual. They need help in keeping mind and spirit strong, so that they can look forward to a future in which they will be able to plan in spite of kidney disease. Not only is the individual a victim of his or her disease (just as is the child), but in the case of parents they often feel responsible and has a self-image as a bad person who had this coming to him or her. As a consequence, the individual drives people away who try to come close and help. Too often, in their pain and confusion, patients/parents do not feel they deserve to be helped, so they let guilt, anger, and self-imposed loneliness makes a bad situation worse. Just when we need others the most, we push them away. A physician with a life-threatening disease wrote: "The patient has been blamed for his illness, has been handed back his questions unopened, and has been left feeling rejected, abandoned. This is classical in long chronic disease." Today, this statement must not be allowed to be anyone's reality.

Support system rationale

I cannot stress enough the need to initiate a process that will allow a person to internalize that which is necessary to live successfully, either as a person with chronic kidney disease or the parent of a child with kidney disease, and in turn facilitate transfer of this knowledge to his or her child. This process will allow the building not only of a therapeutic alliance between the family and the professional, but also a support system that will restore much that has been lost to both the individual with the disease as well as the family – that which is central to truly living, rather than just coping.

The characteristics or formulas that assure psychological survival are communication, control, conviction, clearness of conscience, and an identified life's goal, whether short or long term, and the belief in its attainment. For the young child, it might be going to summer camp, or joining the baseball or swimming team; for the adolescent, it might be college or a career, for the adult it might be maintaining one's employment. Those of us who work with kidney disease patients recognize these as common characteristics in those who function successfully in all aspects of their lives.

A support group need only consist of two people with a common need and the willingness to communicate experience, and thus methods of coping. It need not be a formal group, held regularly, or be problem specific. Just as there are differences in your patients and their families, there are also differences in the type of support they will accept. For some, connection to a formal group setting is of utmost importance; for others, the phone is their lifeline.

Research shows that a solid stable connection to a social group or to humanity in general improves one's resistance to illness because the individual is valuable as a member of the group. This may be one of the reasons why almost all societies emphasize caring for others, being generous to others, and serving them. Perhaps doing so is helpful not only to the entire community, but also to the health of the recipient and donor alike. Thus, it may be true that the isolation one feels during many lengthy hospital stays and living with a chronic disease is not only unhealthy psychologically, but also has an adverse physiological impact.

Physicians know that a sense of independence and self responsibility reflects a healthy psychological adjustment to chronic disease. Dr. Kenneth Terkelson (Rochester, NY) states that "people suffering from a disease may have the expertise they can share with others who have the same disease. We haven't thought of support organizations as a place to turn to help patients. We have been ignoring a valuable resource." Dr. Terkelson also states that "support groups, while 'therapeutic,' are not therapy groups but rather are based on the assumption that people have the ability to cope with their own psychosocial problems if they have enough information, positive role models, additional support and encouragement, and some help in learning to deploy all available resources."

In chronic diseases we deal with tertiary prevention. Although we cannot prevent the disease, in many instances, we can prevent exacerbations and complications by providing information, and by inculcating positive attitudes and the belief that even in a life touched by chronic disease there can be wellness and productivity. Disease-specific support, such as that provided by the Foundation for Children with Atypical HUS, can provide the day-to-day coping mechanisms that give the patient a sense of control – actual, not illusory. Actual control – having concrete things to do to help in the process of recuperation and adjustment – was found to be beneficial to the patient and parent, whereas illusory control – information without a way to act on it – was found to have an adverse effect.

It is true that some things are best incorporated into one's knowledge base through personal experience. It is also true that not everyone can do this alone. There are various ways a parent can cope, and as a former adolescent with inflammatory bowel disease who had the additional pain of watching my own parents suffer as I suffered, I know that a burden shared is a burden lessened.

How, then, does one manage the emotional aspects of the disease? The patient is the one with the physical symptoms. You, the doctor prescribe treatments, medications but what else can you do? You can let them know that they are not alone. You can suggest that they get in touch with support groups and give them the names of other patients and parents in the same situation. You can let them know that your medical treatment is not the sole key to dealing with this disease. The patient, or the parent of the chronically ill child, must be able to renew their own strength to provide strength for themselves and their children. However, parents and children may not recognize the need for a support system. In this case both the physician and the family must recognize that seeking support outside the doctor's office is not an admission of defeat or weakness on the part of the physician, on the part of the patient, or on the part of the parents.

The child's/adolescent's reality

Kristina, a 13-year-old with a chronic disease, writes: "I remember when I was 9 years old; I was going through a very hard time. I thought it was not fair. I remember when my mother told me I had ... disease and how upset she was ... I was scared. I had to take prednisone. It made me get very heavy and a lot of people at school made fun of me. It hurt so badly. The teachers did not understand and wouldn't allow me to go to the bathroom and one day I went in class. Through all of this I needed a friend, someone to talk to who understood."

Laurie K., whose father was diagnosed as having a chronic disease at age 18, is a young woman who herself was found to have the same disease at 17. As both a patient and a family member, she has a unique perspective to share, "The physical drama itself cannot touch us until someone points out its spiritual sense", (Antoine de St. Exupéry, *Wind, Sand and Stars*). Ms. K. points out that, as the "expert," the one to whom the patient and parents turn (or to whom they are dragged, finally, kicking and screaming) in the face of chronic illness, the physician can do far more than talk at those involved. She or he is the initial source of information for the patient and his or her family, and can do more than translate the bewildering medical terms into plain, frightening English.

By the time a child has been diagnosed with a chronic illness, both the child and the parents may be exhausted, confused, angry. There are feelings of helplessness and hopelessness, guilt. Parents want to combat the "thing" that has attacked their child. Medical intervention is indeed an important part of "managing" a disease like chronic or end-stage kidney disease. However, medical management is not enough.

Ms. K. also notes that "once a chronic illness has been diagnosed, there is a feeling of relief; temporarily the symptoms that have disrupted normalcy for the child and the family have finally been labeled. But then comes the same feelings as before. This disease is not curable. It can be managed, but it will never be cured. Parents must be made a vital part of this management team just as the physician prescribes medication to relieve symptoms; parents are providers of emotional comfort. They are the ones on whom the children vent their frustration, their anger, and their tears. But the child is not the only one feeling these things. The parents have the additional burden of their own fears, their own frustrations, and their own tears. To do their best for their child and themselves in order for the family of the chronically ill child not just to manage the illness, but to live, the parents must have somewhere to turn themselves. As the symptoms cause incredible pain and disability for the child, both the child and the parent deal with incredible emotional pain. The physical pain and symptoms can be managed medically, but what about the emotional pain? Just as many diseases cannot be cured, there is no quick cure for the frustration, the hopelessness, helplessness, fear, and anger that go along with chronic disease. Education and support can lessen these feelings and their impact.

Support system

A support group does not have to be a large group of people exposing themselves and their fears to every one else as Ms. Koven's points out. It does not have to be led by a professional or be incorporated. A support group can consist of one or two other people; it can take the form of a book, keeping a journal, or talking to God. When the patient or parent is in the midst of a crisis, or if everyone is worn down by a long struggle with the illness, the patient/parent may not have the energy to get involved in anything formal.

When things are going well, they don't want to be involved in something that reminds them of their ties to the disease and everything awful that goes along with it. Nevertheless, without some support system, formal or informal, one cannot sustain the energy needed to facilitate the healing of mind, body, and spirit of their family unit.

The parent speaks

The words below have been adapted from those of several parents.

A mother of a 5-year-old writes: "Eight months ago our son was found to have atypical HUS. We naturally felt – why us, God? We were told to accept. We had no other choice but to do just that, but we weren't sure what it was we were supposed to accept. We knew nothing of atypical HUS and few others did either. Our doctors treated our son expertly, but they are busy people who sometimes do not realize how ill-informed lay people are; certainly they are too busy to educate whole families, yet whole families are affected.

How often I wished that we could talk to someone about our son's condition who would understand. Then we heard of the Foundation for Children with Atypical HUS. Other people were coping with our same problems. We spoke to other parents and shared concerns and knowledge. We spoke to strangers who were strangers no longer and, best of all we were informed."

To raise a resilient child

Dr. Jules Segal advises that a parent (or professional):

(i) Give youngsters a sense of independence and control over their own destinies. Allow them to make choices as soon as they are able, and show them how they can affect their environments so they will not feel helpless.

(ii) Discourage self blame. Children (and parents) tend to hold themselves responsible for events well beyond their control, and they can be helped to accept the fact that they are not responsible for their illness or the stresses it causes their parents or siblings.

(iii) Do all within the given parameters to allow the child and family to maintain their normal routine.

(iv) Most importantly, communicate. Impress upon the parent that it is vital to open up and that you the professional are there to listen and respond and, if necessary, to act as his or her advocate.

One parent of a child with kidney disease said "When we got together and discussed how to handle teachers, friends, siblings, and other family members, we discovered practical steps we could take to actively address the situation at hand. Our frustration was no longer 'our' frustration."

Support groups serve as a means of channeling tension and frustration that must not be held inside; anxiety turned inward only results in depression. Working as a team, whether it is a team of 2 or 20, pools knowledge and provides direction and self confidence. For example, inflammatory bowel disease, at times, necessitates re-evaluation of previously set goals. That is how we survive life as a whole individual – by being flexible. Only by understanding and hearing this from those who possess a validity that only one who has been there possesses can that message be transmitted. In support groups, the members with validity and empathy say "that it's OK, your feelings are normal, are shared, and can be coped with – have done the same".

Professional lifeline

You are a patient's parent's lifeline, thus your patient's lifeline. By providing a phone number or the name of another parent in your practice who has traveled the chronic disease road for a while longer and, through education and support, has achieved optimal wellness for his or her family unit, you can provide a successful role model, a mentor support. As one physician, himself afflicted with chronic life-threatening disease, put it, "Patient (parent) to patient (parent) support is the life-line to others who have not as yet traveled life's unpredictable road." Those having the best chance of living well with chronic disease, indeed are those with faith in themselves and the physician, connection to a support system, and with the power within. We heard of the National Kidney Foundation. Other people were coping with our same problems. We met other patients and parents, shared concerns and knowledge. We met strangers who were strangers no longer and, best of all, we were informed."

A multidimensional approach to the process of education – true education that facilitates behavior change, and thus adherence to a recommended treatment regimen – can be achieved through the utilization of written material, audiovisual aids, individual counseling, and parent-to-parent and patient-to-patient support. Allowing a parent to view "through that support system looking glass" those parents and children who have succeeded and are whole will allow you to achieve our common goals.

I think back to an incident somewhere in the midst of a 5-month hospitalization. My surgeon came in and found me crying and asked if I was in pain. I replied that I was crying for what could have been and what would never be. It is at that point, as well as at the time your "families" are diagnosed or are in the midst of the current crisis, that they cannot see any positive amidst so much negative. It is education and support that has the power to allow one to tap into that strength within. It is education and support systems – formal and informal – that enable those afflicted with inflammatory bowel disease and other diseases to share methods of coping, thus reinstating the parent and entire family into society as a participant and not an observer.

One parent of a child with a chronic disease said, "When we got together and discussed how to handle teachers, friends, siblings, and other family members, we discovered practical steps we could take to actively address the situation at hand. Our frustration was no longer 'our' frustration." Support groups serve as a means of channeling tension and frustration that must not be held inside; anxiety turned inward becomes depression.

In conclusion

As a person who has lived with a chronic disease for over 40 years, as the mother of four with chronic illnesses, two with atypical HUS, I have learned that central to living well with chronic illness is control. Control can only be gained by educating ourselves about the disease, its complications, treatments and prognosis. We have no control over what disease we have or the level of medical knowledge available at that time. We cannot navigate this effort alone; we must reach out to others living successfully despite the disease.

As a physician you must be confident enough to realize that a patient who feels he/she has no control over his/her life will never be successful in life, will feel victimized by his/her disease. You have the ability to provide your patient with the education and support necessary to foster growth of spirit despite living with the day to day constraints of disease.

We are flesh and bone, mind and body, spirit and soul; you cannot treat one without the other. As a physician you cannot successfully treat the human being in his/her entirety, much less the whole family in its entirety by yourself. You must view yourself as part of a continuum; primary care physician, specialist, nurse, social worker, dietitian, health educator and patient advocacy group. We must all be part of the "team" working together towards the same goal – that of allowing those with chronic illness to engage life with the education and support necessary to be successful in each of our roles – as an individual, family member and member of society.

Index

The PIR-Series
Progress in Inflammation Research

Homepage: http://www.birkhauser.ch

Up-to-date information on the latest developments in the pathology, mechanisms and therapy of inflammatory disease are provided in this monograph series. Areas covered include vascular responses, skin inflammation, pain, neuroinflammation, arthritis cartilage and bone, airways inflammation and asthma, allergy, cytokines and inflammatory mediators, cell signalling, and recent advances in drug therapy. Each volume is edited by acknowledged experts providing succinct overviews on specific topics intended to inform and explain. The series is of interest to academic and industrial biomedical researchers, drug development personnel and rheumatologists, allergists, pathologists, dermatologists and other clinicians requiring regular scientific updates.

Available volumes:

T Cells in Arthritis, P. Miossec, W. van den Berg, G. Firestein (Editors), 1998
Chemokines and Skin, E. Kownatzki, J. Norgauer (Editors), 1998
Medicinal Fatty Acids, J. Kremer (Editor), 1998
Inducible Enzymes in the Inflammatory Response,
D.A. Willoughby, A. Tomlinson (Editors), 1999
Cytokines in Severe Sepsis and Septic Shock, H. Redl, G. Schlag (Editors), 1999
Fatty Acids and Inflammatory Skin Diseases, J.-M. Schröder (Editor), 1999
Immunomodulatory Agents from Plants, H. Wagner (Editor), 1999
Cytokines and Pain, L. Watkins, S. Maier (Editors), 1999
In Vivo *Models of Inflammation*, D. Morgan, L. Marshall (Editors), 1999
Pain and Neurogenic Inflammation, S.D. Brain, P. Moore (Editors), 1999
Anti-Inflammatory Drugs in Asthma, A.P. Sampson, M.K. Church (Editors), 1999
Novel Inhibitors of Leukotrienes, G. Folco, B. Samuelsson, R.C. Murphy (Editors), 1999
Vascular Adhesion Molecules and Inflammation, J.D. Pearson (Editor), 1999
Metalloproteinases as Targets for Anti-Inflammatory Drugs,
K.M.K. Bottomley, D. Bradshaw, J.S. Nixon (Editors), 1999
Free Radicals and Inflammation, P.G. Winyard, D.R. Blake, C.H. Evans (Editors), 1999
Gene Therapy in Inflammatory Diseases, C.H. Evans, P. Robbins (Editors), 2000
New Cytokines as Potential Drugs, S. K. Narula, R. Coffmann (Editors), 2000
High Throughput Screening for Novel Anti-inflammatories, M. Kahn (Editor), 2000
Immunology and Drug Therapy of Atopic Skin Diseases,
C.A.F. Bruijnzeel-Komen, E.F. Knol (Editors), 2000
Novel Cytokine Inhibitors, G.A. Higgs, B. Henderson (Editors), 2000
Inflammatory Processes. Molecular Mechanisms and Therapeutic Opportunities,
L.G. Letts, D.W. Morgan (Editors), 2000

Cellular Mechanisms in Airways Inflammation, C. Page, K. Banner, D. Spina (Editors), 2000
Inflammatory and Infectious Basis of Atherosclerosis, J.L. Mehta (Editor), 2001
Muscarinic Receptors in Airways Diseases, J. Zaagsma, H. Meurs, A.F. Roffel (Editors), 2001
TGF-β and Related Cytokines in Inflammation, S.N. Breit, S. Wahl (Editors), 2001
Nitric Oxide and Inflammation, D. Salvemini, T.R. Billiar, Y. Vodovotz (Editors), 2001
Neuroinflammatory Mechanisms in Alzheimer's Disease. Basic and Clinical Research, J. Rogers (Editor), 2001
Disease-modifying Therapy in Vasculitides, C.G.M. Kallenberg, J.W. Cohen Tervaert (Editors), 2001
Inflammation and Stroke, G.Z. Feuerstein (Editor), 2001
NMDA Antagonists as Potential Analgesic Drugs, D.J.S. Sirinathsinghji, R.G. Hill (Editors), 2002
Migraine: A Neuroinflammatory Disease? E.L.H. Spierings, M. Sanchez del Rio (Editors), 2002
Mechanisms and Mediators of Neuropathic pain, A.B. Malmberg, S.R. Chaplan (Editors), 2002
Bone Morphogenetic Proteins. From Laboratory to Clinical Practice, S. Vukicevic, K.T. Sampath (Editors), 2002
The Hereditary Basis of Allergic Diseases, J. Holloway, S. Holgate (Editors), 2002
Inflammation and Cardiac Diseases, G.Z. Feuerstein, P. Libby, D.L. Mann (Editors), 2003
Mind over Matter – Regulation of Peripheral Inflammation by the CNS, M. Schäfer, C. Stein (Editors), 2003
Heat Shock Proteins and Inflammation, W. van Eden (Editor), 2003
Pharmacotherapy of Gastrointestinal Inflammation, A. Guglietta (Editor), 2004
Arachidonate Remodeling and Inflammation, A.N. Fonteh, R.L. Wykle (Editors), 2004
Recent Advances in Pathophysiology of COPD, P.J. Barnes, T.T. Hansel (Editors), 2004
Cytokines and Joint Injury, W.B. van den Berg, P. Miossec (Editors), 2004
Cancer and Inflammation, D.W. Morgan, U. Forssmann, M.T. Nakada (Editors), 2004
Bone Morphogenetic Proteins: Bone Regeneration and Beyond, S. Vukicevic, K.T. Sampath (Editors), 2004
Antibiotics as Anti-Inflammatory and Immunomodulatory Agents, B.K. Rubin, J. Tamaoki (Editors), 2005
Antirheumatic Therapy: Actions and Outcomes, R.O. Day, D.E. Furst, P.L.C.M. van Riel, B. Bresnihan (Editors), 2005
Regulatory T-Cells in Inflammation, L. Taams, A.N. Akbar, M.H.M Wauben (Editors), 2005
Sodium Channels, Pain, and Analgesia, K. Coward, M. Baker (Editors), 2005
Turning up the Heat on Pain: TRPV1 Receptors in Pain and Inflammation, A.B Malmberg, K.R. Bley (Editors), 2005
The NPY Family of Peptides in Immune Disorders, Inflammation, Angiogenesis and Cancer, Zofia Zukowska, Giora Z. Feuerstein (Editors), 2005